Encyclopaedia of
Nanotechnology

Encyclopaedia of **Nanotechnology**

Vol. 10 : Nanosnsing

Dr. Rita Khare

AGROTECH PRESS

JAIPUR ■ NEW DELHI

AGROTECH PRESS

Regd. Off. "Bony Residency", Gate No. 2
Opp. Tilak Public School, Vishweswariya Nagar,
Gopalpura Road, Jaipur - 302018, India
Ph: 0141 - 2761280, Telefax: 2761381
E-mail: agrotechpress@gmail.com

Delhi Office: 102, 1st Floor, Satyam House,
4327/3, Ansari Road, Darya Ganj,
New Delhi - 110002, India
Ph: 011-45652440
visit us: www.agrotechpress.com

Distributed by :
KEYA INTERNATIONAL PRESS LLP
E-mail: keyainternationalpressllp@gmail.com

Published in India, 2024

ISBN : 9789384568566 (10 Vol. Set)

Typeset by Agrotech Press

Printed and bound in India by Replika Press Pvt. Ltd.

Contents

Preface

Nanoscience and Nanotechnologies have grown increasingly fast for the last 20 years - after a somehow slow start - with an exceptional impact upon understanding of nature, development of science and related applications. Several new materials have been built and the possibility of tailoring their properties for particular purposes has opened unexpected perspectives in a multidisciplinary scenario. Furthermore, economy has also been deeply involved in this effort as a consequence of the fact that a significant amount of money has been invested in new enterprises. Nanoscience and Nanotechnologies are developing at a very fast pace. It is then important to provide expert scientists with the possibility of reviewing and updating some of the most significant features of nanostructures for a better understanding of their scientific foundations in order to put a firm basis for future developments. Nanostructures and nnosensing applications provide a very effective approach as they require a strict interaction between science and technology leading to a high degree of usage.

Nanosensors are any biological, chemical, or surgical sensory points used to convey information about nanoparticles to the macroscopic world. Nanosensors are inherently more sensitive than any other kind of sensor, making them a future choice where lives are at stake. In addition, their small size and potentially low cost means that they can be widely deployed. Their use mainly include various medicinal purposes and as gateways to building other nanoproducts, such as computer chips that work at the nanoscale and nanorobots. Modern research is directed at developing new types of sensors with nanometer dimensions that exploit nanoscale physical, electronic, optical and/or chemical effects. Also to engineer the surfaces and interfaces of nanoscale sensors in order to optimize sensor performance as well as to study fundamental processes related to molecular adsorption and binding. The advancement in detecting and sensing different biological and chemical species with increased capacity and accuracy may transform societal mechanisms that were originally designed on uncertainty and imprecise information. For example, the ability to measure extremely low amounts of air pollutants or toxic materials in water raises questions and dilemmas of risk thresholds especially if the advancement of such technologies outpaces the ability of the public to respond. As such, nanosensors are expected to lead to revolutionary applications, including early disease detection that can result in faster treatments and better outcomes, as well as the early and accurate detection of environmental pollutants, contaminants, and even biological or chemical weapons. Due to the diverse nature of these potential applications, nanosensors are expected to impact multiple sectors of the economy, including the healthcare, pharmaceutical, agricultural, food, environmental, consumer products, and defence sectors.

Medicinal uses of nanosensors mainly revolve around the potential of nanosensors to accurately identify particular cells or places in the body in need. By measuring changes in volume, concentration, displacement and velocity, gravitational, electrical, and magnetic forces, pressure, or temperature of cells in a body, nanosensors may be able to distinguish between and recognize certain cells, most notably those of cancer, at the molecular level in order to deliver medicine or monitor development to specific places in the body. In addition, they may be able to detect macroscopic variations from outside the body and communicate these changes to other nanoproducts working within the body. One example of nanosensors involves using the fluorescence properties of cadmium selenide quantum dots as sensors to uncover tumors within the body. By injecting a body with these quantum dots, a doctor could see where a tumour or cancer cell was by finding the injected quantum dots, an easy process because of their fluorescence. Other projected products most commonly involve using nanosensors to build smaller integrated circuits, as well as incorporating them into various other commodities made using other forms of nanotechnology for use in a variety of situations including transportation, communication, improvements in structural integrity, and robotics. Nanosensors may also eventually be valuable as more accurate monitors of material states for use in systems where size and weight are constrained, such as in satellites and other aeronautic machines.

Though nanosensor technology is a relatively new field, global projections for sales of products incorporating nanosensors range from a few billions in the next decade. They will likely be included in most modern circuitry used in advanced computing systems, since their potential to provide the link between other forms of nanotechnology and the macroscopic world allows developers to fully exploit the potential of nanotechnology to miniaturize computer chips while vastly expanding their storage potential. A complete and authoritative guide to nanosensors, this book offers up-to-date information on the fabrication, properties, and operating mechanisms of these fast and reliable sensors. It addresses progress in the field, fundamental issues and challenges facing researchers, and prospects for future development.

Dr. Rita Khare

1

Introduction

A sensor is a converter that measures a physical quantity and converts it into a signal which can be read by an observer or by an (today mostly electronic) instrument. For example, a mercury-in-glass thermometre converts the measured temperature into expansion and contraction of a liquid which can be read on a calibrated glass tube. A thermocouple converts temperature to an output voltage which can be read by a voltmetre. For accuracy, most sensors are calibrated against known standards.

Sensors are used in everyday objects such as touch-sensitive elevator buttons (tactile sensor) and lamps which dim or brighten by touching the base. There are also innumerable applications for sensors of which most people are never aware. Applications include cars, machines, aerospace, medicine, manufacturing and robotics.

A sensor is a device, which responds to an input quantity by generating a functionally related output usually in the form of an electrical or optical signal. A sensor's sensitivity indicates how much the sensor's output changes when the measured quantity changes. For instance, if the mercury in a thermometre moves 1 cm when the temperature changes by 1 °C, the sensitivity is 1 cm/°C (it is basically the slope Dy/Dx assuming a linear characteristic). Sensors that measure very small changes must have very high sensitivities. Sensors also have an impact on what they measure; for instance, a room temperature thermometre inserted into a hot cup of liquid cools the liquid while the liquid heats the thermometre. Sensors need to be designed to have a small effect on what is measured; making the sensor smaller often improves this and may introduce other advantages. Technological progress allows more and more sensors to be manufactured on a microscopic scale as microsensors using MEMS technology. In most cases, a microsensor reaches a significantly higher speed and sensitivity compared with macroscopic approaches.

Nanosensor

Nanosensors are any biological, chemical, or surgical sensory points used to convey information about nanoparticles to the macroscopic world. Their use mainly include various medicinal purposes and as gateways to building other nanoproducts, such as computer chips that work at the nanoscale and nanorobots. Presently, there are several ways proposed to make nanosensors, including top-down lithography, bottom-up assembly, and molecular self-assembly.

Predicted Applications

Medicinal uses of nanosensors mainly revolve around the potential of nanosensors to accurately identify particular cells or places in the body in need. By measuring changes in volume, concentration, displacement and velocity, gravitational, electrical, and magnetic forces, pressure, or temperature of cells in a body, nanosensors may be able to distinguish between and recognise certain cells, most notably those of cancer, at the molecular level in order to deliver medicine or monitor development to specific places in the body. In addition, they may be able to detect macroscopic variations from outside the body and communicate these changes to other nanoproducts working within the body.

One example of nanosensors involves using the fluorescence properties of cadmium selenide quantum dots as sensors to uncover tumors within the body. By injecting a body with these quantum dots, a doctor could see where a tumor or cancer cell was by finding the injected quantum dots, an easy process because of their fluorescence. Developed nanosensor quantum dots would be specifically constructed to find only the particular cell for which the body was at risk. A downside to the cadmium selenide dots, however, is that they are highly toxic to the body. As a result, researchers are working on developing alternate dots made out of a different, less toxic material while still retaining some of the fluorescence properties. In particular, they have been investigating the particular benefits of zinc sulfide quantum dots which, though they are not quite as fluorescent as cadmium selenide, can be augmented with other metals including manganese and various lanthanide elements. In addition, these newer quantum dots become more fluorescent when they bond to their target cells. (Quantum) Potential predicted functions may also include sensors used to detect specific DNA in order to recognise explicit genetic defects, especially for individuals at high-risk and implanted sensors that can automatically detect glucose levels for diabetic subjects more simply than current detectors. DNA can also serve as sacrificial layer for manufacturing CMOS IC, integrating a nanodevice with sensing capabilities. Therefore, using proteomic patterns and new hybrid materials, nanobiosensors can also be used to enable components configured into a hybrid semiconductor substrate as part of the circuit assembly. The development and miniaturization of nanobiosensors should provide interesting new opportunities.

Other projected products most commonly involve using nanosensors to build smaller integrated circuits, as well as incorporating them into various other commodities made using other forms of nanotechnology for use in a variety of situations including transportation, communication, improvements in structural integrity, and robotics. Nanosensors may also eventually be valuable as more accurate monitors of material states for use in systems where size and weight are constrained, such as in satellites and other aeronautic machines.

Existing Nanosensors

Currently, the most common mass-produced functioning nanosensors exist in the biological world as natural receptors of outside stimulation. For instance, sense of smell, especially in animals in which it is particularly strong, such as dogs, functions using receptors that sense nanosized molecules. Certain plants, too, use nanosensors to detect sunlight; various fish use nanosensors to detect minuscule vibrations in the surrounding water; and many insects detect sex pheromones using nanosensors.

One of the first working examples of a synthetic nanosensor was built by researchers at the Georgia Institute of Technology in 1999. It involved attaching a single particle onto the end of a carbon nanotube and measuring the vibrational frequency of the nanotube both with and without the particle. The discrepancy between the two frequencies allowed the researchers to measure the mass of the attached particle.

Chemical sensors, too, have been built using nanotubes to detect various properties of gaseous molecules. Carbon nanotubes have been used to sense ionization of gaseous molecules while nanotubes made out of titanium have been employed to detect atmospheric concentrations of hydrogen at the molecular level. Many of these involve a system by which nanosensors are built to have a specific pocket for another molecule. When that particular molecule, and only that specific molecule, fits into the nanosensor, and light is shone upon the nanosensor, it will reflect different wavelengths of light and, thus, be a different colour. In a similar fashion, Flood et al. have shown that supramolecular host-guest chemistry offers quantitative sensing using Raman scattered light as well as SERS.

Production Methods

There are currently several hypothesized ways to produce nanosensors. Top-down lithography is the manner in which most integrated circuits are now made. It involves starting out with a larger block of some material and carving out the desired form. These carved out devices, notably put to use in specific microelectromechanical systems used as microsensors, generally only reach the micro size, but the most recent of these have begun to incorporate nanosized components.

Another way to produce nanosensors is through the bottom-up method, which involves assembling the sensors out of even more minuscule components, most likely individual atoms or molecules. This would involve moving atoms of a particular substance one by one into particular positions which, though it has been achieved in laboratory tests using tools such as atomic force microscopes, is still a significant difficulty, especially to do en masse, both for logistic reasons as well as economic ones. Most likely, this process would be used mainly for building starter molecules for self-assembling sensors.

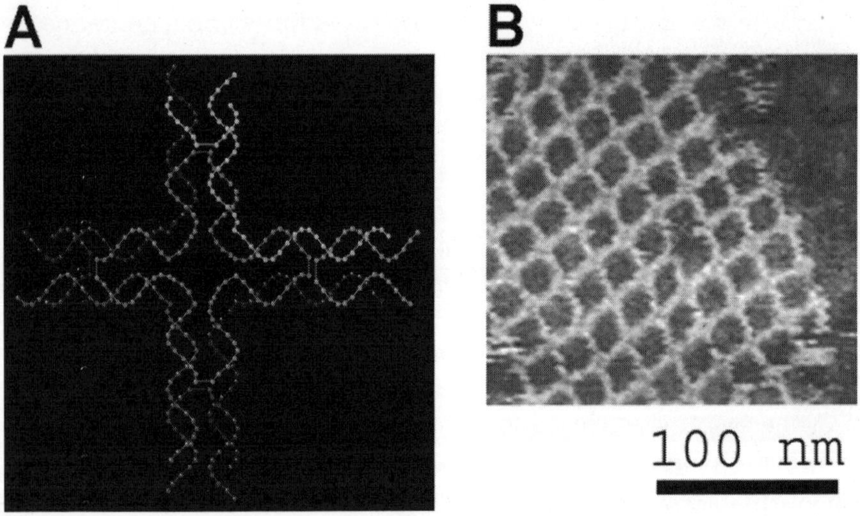

Figure: *(A) An example of a DNA molecule used as a starter for larger self-assembly. (B) An atomic force microscope image of a self-assembled DNA nanogrid. Individual DNA tiles self-assemble into a highly ordered periodic two-dimensional DNA nanogrid.*

The third way, which promises far faster results, involves self-assembly, or "growing" particular nanostructures to be used as sensors. This most often entails one of two types of assembly. The first involves using a piece of some previously created or naturally formed nanostructure and immersing it in free atoms of its own kind. After a given period, the structure, having an irregular surface that would make it prone to attracting more molecules as a continuation of its current pattern, would capture some of the free atoms and continue to form more of itself to make larger components of nanosensors.

The second type of self-assembly starts with an already complete set of components that would automatically assemble themselves into a finished product. Though this has been so far successful only in assembling computer chips at the micro size, researchers hope to eventually be able to do it at the nanometre size for multiple products, including nanosensors. Accurately being able to reproduce this effect for a desired sensor in a laboratory would imply that scientists could manufacture nanosensors much more quickly and potentially far more cheaply by letting numerous molecules assemble themselves with little or no outside influence, rather than having to manually assemble each sensor.

Economic Impacts

Though nanosensor technology is a relatively new field, global projections for sales of products incorporating nanosensors range from \$0.6 billion to \$2.7 billion in the next three to four years. They will likely be included in most modern circuitry used in advanced computing systems, since their potential to provide the link between other forms of nanotechnology and the macroscopic world allows developers to fully exploit the potential of nanotechnology to miniaturize computer chips while vastly expanding their storage potential.

First, however, nanosensor developers must overcome the present high costs of production in order to become worthwhile for implementation in consumer products. Additionally, nanosensor reliability is not yet suitable for widespread use, and, because of their scarcity, nanosensors have yet to be marketed and implemented outside of research facilities. Consequently, nanosensors have yet to be made compatible with most consumer technologies for which they have been projected to eventually enhance.

Social Impacts

Ethical and social impacts are harder to define and sort as good or bad compared to health and environmental impacts. The advancement in detecting and sensing different biological and chemical species with increased capacity and accuracy may transform societal mechanisms that were originally designed on uncertainty and imprecise information. For example, the ability to measure extremely low amounts of air pollutants or toxic materials in water raises questions and dilemmas of risk thresholds especially if the advancement of such technologies outpaces the ability of the public to respond. As another example, medical sensors will not only help in diagnoses and treatment but may also predict the future profile of an individual. This will add to the information used by health insurance companies to grant or deny coverage.

Other social issues resulting from the widespread use of nanosensors and surveillance devices include privacy invasion and security issues.

Sensor Characteristics and Physical Effects

Sensors provide an output signal in response to a physical, chemical, or biological measure and typically require an electrical input to operate them; therefore, they tend to be characterized in specifications or datasheets in much the same way as electronic devices are characterized. To truly understand sensors, and how sensors that measure the same measure and differ, it is necessary to understand sensor performance characteristics. Unfortunately, there is no standard definition for many of these characteristics and different parts of the sensor community have different names for these characteristics, depending on the sensing domain. This confusion is further compounded by manufacturers publishing an abundance of performance characteristics, thus making it difficult for the potential user to identify those which are relevant to their application, and how they should be applied.

Photoelectric Effect

The photoelectric effect is the observation that many metals emit electrons when light shines upon them. Electrons emitted in this manner may be called *photoelectrons*.

According to classical electromagnetic theory, this effect can be attributed to the transfer of energy from the light to an electron in the metal. From this perspective, an alteration in either the amplitude or wavelength of light would induce changes in the rate of emission of electrons from the metal. Furthermore, according to this theory, a sufficiently dim light would be expected to show a lag time between the initial shining of its light and the subsequent emission of an electron. However, the experimental results did not correlate with either of the two predictions made by this theory.

Instead, as it turns out, electrons are only dislodged by the photoelectric effect if light reaches or exceeds a threshold frequency, below which no electrons can be emitted from the metal regardless of the amplitude and temporal length of exposure of light. To make sense of the fact that light can eject electrons even if its intensity is low, Albert Einstein proposed that a beam of light is not a wave propagating through space, but rather a collection of discrete wave packets (photons), each with energy hf. This shed light on Max Planck's previous discovery of the Planck relation ($E = hf$) linking energy (E) and frequency (f) as arising from quantization of energy. The factor h is known as the Planck constant.

In 1887, Heinrich Hertz discovered that electrodes illuminated with ultraviolet light create electric sparks more easily. In 1905 Albert Einstein published a paper that explained experimental data from the photoelectric effect as being the result of light energy being carried in discrete quantized packets. This discovery led to the quantum revolution. Einstein was awarded the Nobel Prize in 1921 for "his discovery of the law of the photoelectric effect".

The photoelectric effect requires photons with energies from a few electronvolts to over 1 MeV in elements with a high atomic number. Study of the photoelectric effect led to important steps in understanding the quantum nature of light and electrons and influenced the formation of the concept of wave–particle duality. Other phenomena where light affects the movement of electric charges include the photoconductive effect (also known as photoconductivity or photoresistivity), the photovoltaic effect, and the photoelectrochemical effect.

Emission Mechanism

The photons of a light beam have a characteristic energy proportional to the frequency of the light. In the photoemission process, if an electron within some material absorbs the energy of one photon and acquires more energy than the work function (the electron binding energy) of the material, it is ejected. If the photon energy is too low, the electron is unable to escape the material. Since an increase in the intensity of low-frequency light will only increase the number of low-energy photons sent over a given interval of time, this change in intensity will not create any single photon with enough energy to dislodge an electron. Thus, the energy of the emitted electrons does not depend on the intensity of the incoming light, but only on the energy (equivalently frequency) of the individual photons. It is an interaction between the incident photon and the outermost electrons.

Electrons can absorb energy from photons when irradiated, but they usually follow an "all or nothing" principle. All of the energy from one photon must be absorbed and used to liberate one electron from atomic binding, or else the energy is re-emitted. If the photon energy is absorbed, some of the energy liberates the electron from the atom, and the rest contributes to the electron's kinetic energy as a free particle.

Experimental Observations of Photoelectric Emission

The theory of the photoelectric effect must explain the experimental observations of the emission of electrons from an illuminated metal surface.

For a given metal, there exists a certain minimum frequency of incident radiation below which no photoelectrons are emitted. This frequency is called the *threshold frequency*. Increasing the frequency of the incident beam, keeping the number of incident photons fixed (this would result in a proportionate increase in energy) increases the maximum kinetic energy of the photoelectrons emitted. Thus the stopping voltage increases. The number of electrons also changes because the probability that each photon results in an emitted electron is a function of photon energy. If the intensity of the incident radiation is increased, there is no effect on the kinetic energies of the photoelectrons.

Above the threshold frequency, the maximum kinetic energy of the emitted photoelectron depends on the frequency of the incident light, but is independent of the intensity of the incident light so long as the latter is not too high.

For a given metal and frequency of incident radiation, the rate at which photoelectrons are ejected is directly proportional to the intensity of the incident light. Increase in intensity of incident beam (keeping the frequency fixed) increases the magnitude of the photoelectric current, though stopping voltage remains the same.

The time lag between the incidence of radiation and the emission of a photoelectron is very small, less than 10^{-9} second.

The direction of distribution of emitted electrons peaks in the direction of polarization (the direction of the electric field) of the incident light, if it is linearly polarized.

Mathematical Description

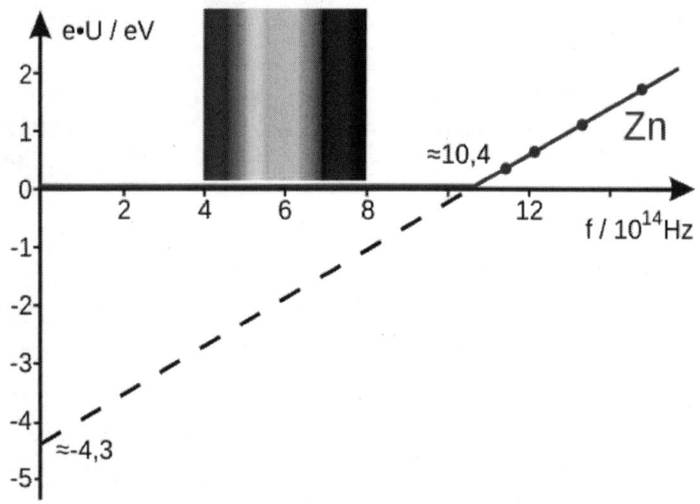

Figure: *Diagram of the maximum kinetic energy as a function of the frequency of light on zinc*

The maximum kinetic energy $K_{\max}$ of an ejected electron is given by

$$K_{\max} = hf - \varphi,$$

where h is the Planck constant and f is the frequency of the incident photon. The term φ is the work function (sometimes denoted W, or ϕ), which gives the minimum energy required to remove a delocalised electron from the surface of the metal. The work function satisfies

$$\varphi = hf_0,$$

where f_0 is the threshold frequency for the metal. The maximum kinetic energy of an ejected electron is then

$$K_{max} = h(f - f_0).$$

Kinetic energy is positive, so we must have $f > f_0$ for the photoelectric effect to occur.

Stopping Potential

The relation between current and applied voltage illustrates the nature of the photoelectric effect. For discussion, a light source illuminates a plate P, and another plate electrode Q collects any emitted electrons. We vary the potential between P and Q and measure the current flowing in the external circuit between the two plates.

If the frequency and the intensity of the incident radiation are fixed, the photoelectric current increases gradually with an increase in positive potential on collector electrode until all the photoelectrons emitted are collected. The photoelectric current attains a saturation value and does not increase further for any increase in the positive potential. The saturation current depends on the intensity of illumination, but not its wavelength.

If we apply a negative potential to plate Q with respect to plate P and gradually increase it, the photoelectric current decreases until it is zero, at a certain negative potential on plate Q. The minimum negative potential given to plate Q at which the photoelectric current becomes zero is called stopping potential or cut off potential

i. For the given frequency of incident radiation, the stopping potential is independent of its intensity.

ii. For a given frequency of the incident radiation, the stopping potential Vo is related to the maximum kinetic energy of the photoelectron that is just stopped from reaching plate Q. If m is the mass and v_{max} is the maximum velocity of photoelectron emitted, then

$$K_{max} = \frac{1}{2}mv_{max}^2$$

If q_e is the charge on the electron and V_0 is the stopping potential, then the work done by the retarding potential in stopping the electron $= q_eV_0$, which gives

$$\frac{1}{2}mv_{max}^2 = q_eV_0$$

The above relation shows that the maximum velocity of the emitted photoelectron is independent of the intensity of the incident light. Hence,

$$K_{max} = q_e V_0$$

The stopping voltage varies linearly with frequency of light, but depends on the type of material. For any particular material, there is a threshold frequency that must be exceeded, independent of light intensity, to observe any electron emission.

Three-Step Model

In the X-ray regime, the photoelectric effect in crystalline material is often decomposed into three steps:

1. Inner photoelectric effect. The hole left behind can give rise to auger effect, which is visible even when the electron does not leave the material. In molecular solids phonons are excited in this step and may be visible as lines in the final electron energy. The inner photoeffect has to be dipole allowed. The transition rules for atoms translate via the tight-binding model onto the crystal. They are similar in geometry to plasma oscillations in that they have to be transversal.
2. Ballistic transport of half of the electrons to the surface. Some electrons are scattered.
3. Electrons escape from the material at the surface.

In the three-step model, an electron can take multiple paths through these three steps. All paths can interfere in the sense of the path integral formulation. For surface states and molecules the three-step model does still make some sense as even most atoms have multiple electrons which can scatter the one electron leaving.

History

When a surface is exposed to electromagnetic radiation above a certain threshold frequency (typically visible light for alkali metals, near ultraviolet for other metals, and extreme ultraviolet for non-metals), the radiation is absorbed and electrons are emitted. Light, and especially ultra-violet light, discharges negatively electrified bodies with the production of rays of the same nature as cathode rays. Under certain circumstances it can directly ionize gases. The first of these phenomena was discovered by Hertz and Hallwachs in 1887. The second was announced first by Philipp Lenard in 1900.

The ultra-violet light to produce these effects may be obtained from an arc lamp, or by burning magnesium, or by sparking with an induction coil between zinc or cadmium terminals, the light from which is very rich in ultraviolet rays. Sunlight is not rich in ultra-violet rays, as these have been absorbed by the atmosphere, and it does not produce nearly so large an effect as the arc-light. Many substances besides metals discharge negative electricity under the action of ultraviolet light: lists of these substances will be found in papers by G. C. Schmidt and O. Knoblauch.

19th Century

In 1839, Alexandre Edmond Becquerel discovered the photovoltaic effect while studying the effect of light on electrolytic cells. Though not equivalent to the photoelectric effect, his work on photovoltaics was instrumental in showing a strong relationship between light and electronic properties of materials. In 1873, Willoughby Smith discovered photoconductivity in selenium while testing the metal for its high resistance properties in conjunction with his work involving submarine telegraph cables.

Johann Elster (1854–1920) and Hans Geitel (1855–1923), students in Heidelberg, developed the first practical photoelectric cells that could be used to measure the intensity of light. Elster and Geitel had investigated with great success the effects produced by light on electrified bodies.

Uses and Effects

Photomultipliers: These are extremely light-sensitive vacuum tubes with a photocathode coated onto part (an end or side) of the inside of the envelope. The photocathode contains combinations of materials such as caesium, rubidium and antimony specially selected to provide a low work function, so when illuminated even by very low levels of light, the photocathode readily releases electrons. By means of a series of electrodes (dynodes) at ever-higher potentials, these electrons are accelerated and substantially increased in number through secondary emission to provide a readily detectable output current. Photomultipliers are still commonly used wherever low levels of light must be detected.

Image Sensors: Video camera tubes in the early days of television used the photoelectric effect, for example, Philo Farnsworth's "Image dissector" used a screen charged by the photoelectric effect to transform an optical image into a scanned electronic signal.

Gold-Leaf Electroscope

Gold-leaf electroscopes are designed to detect static electricity. Charge placed on the metal cap spreads to the stem and the gold leaf of the electroscope. Because they then have the same charge, the stem and leaf repel each other. This will cause the leaf to bend away from the stem.

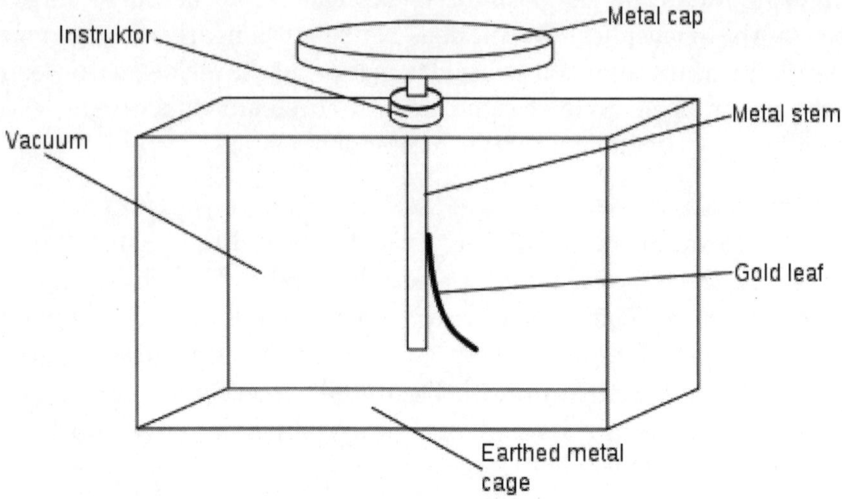

***Figure:** The gold leaf electroscope*

The electroscope is an important tool in illustrating the photoelectric effect. For example, if the electroscope is negatively charged throughout, there is an excess of electrons and the leaf is separated from the stem. If high-frequency light shines on the cap, the electroscope discharges and the leaf will fall limp. This is because the frequency of the light shining on the cap is above the cap's threshold frequency. The photons in the light have enough energy to liberate electrons from the cap, reducing its negative charge. This will discharge a negatively charged electroscope and further charge a positive electroscope. However, if the electromagnetic radiation hitting the metal cap does not have a high enough frequency (its frequency is below the threshold value for the cap), then the leaf will never discharge, no matter how long one shines the low-frequency light at the cap.

Photoelectron Spectroscopy

Since the energy of the photoelectrons emitted is exactly the energy of the incident photon minus the material's work function or binding energy, the work function of a sample can be determined by bombarding it with a monochromatic X-ray source or UV source, and measuring the kinetic energy distribution of the electrons emitted.

Photoelectron spectroscopy is usually done in a high-vacuum environment, since the electrons would be scattered by gas molecules if they were present. However, some companies are now selling products that allow photoemission in air. The light source can be a laser, a discharge tube, or a synchrotron radiation source.

The concentric hemispherical analyser (CHA) is a typical electron energy analyzer, and uses an electric field to change the directions of incident electrons, depending on their kinetic energies. For every element and core (atomic orbital) there will be a different binding energy. The many electrons created from each of these combinations will show up as spikes in the analyzer output, and these can be used to determine the elemental composition of the sample.

Spacecraft

The photoelectric effect will cause spacecraft exposed to sunlight to develop a positive charge. This can be a major problem, as other parts of the spacecraft in shadow develop a negative charge from nearby plasma, and the imbalance can discharge through delicate electrical components. The static charge created by the photoelectric effect is self-limiting, though, because a more highly charged object gives up its electrons less easily.

Moon Dust

Light from the sun hitting lunar dust causes it to become charged through the photoelectric effect. The charged dust then repels itself and lifts off the surface of the Moon by electrostatic levitation. This manifests itself almost like an "atmosphere of dust", visible as a thin haze and blurring of distant features, and visible as a dim glow after the sun has set. This was first photographed by the Surveyor program probes in the 1960s. It is thought that the smallest particles are repelled up to kilometres high, and that the particles move in "fountains" as they charge and discharge.

Night Vision Devices

Photons hitting a thin film of alkali metal or semiconductor material such as gallium arsenide in an image intensifier tube cause the ejection of photoelectrons due to the photoelectric effect. These are accelerated by an electrostatic field where they strike a phosphor coated screen, converting the electrons back into photons. Intensification of the signal is achieved either through acceleration of the electrons or by increasing the number of electrons through secondary emissions, such as with a Micro-channel plate. Sometimes a combination of both methods is used. Additional kinetic energy is required to move an electron out of the conduction band and into the vacuum level. This is known as the electron affinity of the photocathode and is another barrier to photoemission other than the forbidden band, explained by the band gap model. Some materials such as Gallium Arsenide have an effective electron affinity that is below the level of the conduction band. In these materials, electrons that move to the conduction band are all of sufficient energy to be emitted from the material and as such, the film that absorbs photons can be quite thick. These materials are known as negative electron affinity materials.

Cross Section

The photoelectric effect is one interaction mechanism between photons and atoms. It is one of 12 theoretically possible interactions.

At the high photon energies comparable to the electron rest energy of 511 keV, Compton scattering, another process, may take place. Above twice this (1.022 MeV) pair production may take place. Compton scattering and pair production are examples of two other competing mechanisms.

Indeed, even if the photoelectric effect is the favoured reaction for a particular single-photon bound-electron interaction, the result is also subject to statistical processes and is not guaranteed, albeit the photon has certainly disappeared and a bound electron has been excited (usually K or L shell electrons at gamma ray energies). The probability of the photoelectric effect occurring is measured by the cross section of interaction, σ. This has been found to be a function of the atomic number of the target atom and photon energy. A crude approximation, for photon energies above the highest atomic binding energy, is given by:

$$\sigma = \text{constant} \cdot \frac{Z^n}{E^3}$$

Here Z is atomic number and n is a number which varies between 4 and 5. (At lower photon energies a characteristic structure with edges appears, K edge, L edges, M edges, etc.) The obvious interpretation follows that the photoelectric effect rapidly decreases in significance, in the gamma ray region of the spectrum, with increasing photon energy, and that photoelectric effect increases steeply with atomic number. The corollary is that high-Z materials make good gamma-ray shields, which is the principal reason that lead (Z = 82) is a preferred and ubiquitous gamma radiation shield.

Electron Microscopy

An electron microscope (EM) is a type of microscope that uses an electron beam to illuminate a specimen and produce a magnified image.

An EM has greater resolving power than a light microscope and can reveal the structure of smaller objects because electrons have wavelengths about 100,000 times shorter than visible light photons. They can achieve better than 50 pm resolution and magnifications of up to about 10,000,000x whereas ordinary, non-confocal light microscopes are limited by diffraction to about 200 nm resolution and useful magnifications below 2000x.

The electron microscope uses electrostatic and electromagnetic lenses to control the electron beam and focus it to form an image. These electron optical lenses are analogous to the glass lenses of a light optical microscope.

Electron microscopes are used to investigate the ultrastructure of a wide range of biological and inorganic specimens including microorganisms, cells, large molecules, biopsy samples, metals, and crystals. Industrially, the electron microscope is often used for quality control and failure analysis. Modern electron microscopes produce electron micrographs, using specialised digital cameras or frame grabbers to capture the image.

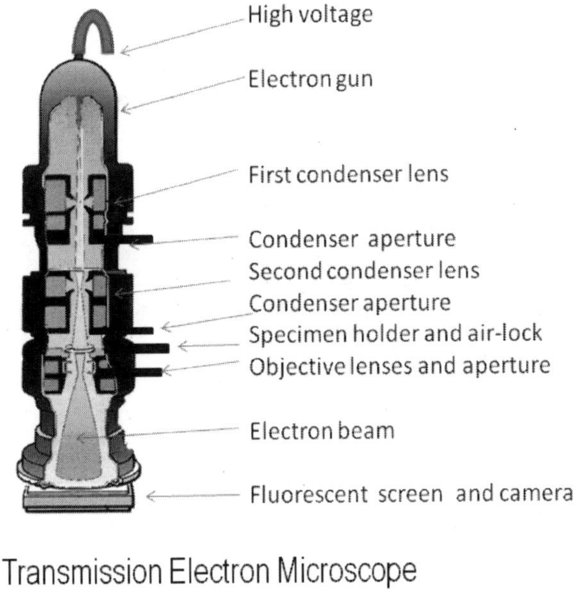

Figure: *The first electromagnetic lens was developed in 1926 by Hans Busch.*

According to Dennis Gabor, the physicist Leσ Szilárd tried in 1928 to convince Busch to build an electron microscope, for which he had filed a patent.

The German physicist Ernst Ruska and the electrical engineer Max Knoll constructed the prototype electron microscope in 1931, capable of four-hundred-power magnification; the apparatus was the first demonstration of the principles of electron microscopy. Two years later, in 1933, Ruska built an electron microscope that exceeded the resolution attainable with an optical (light) microscope. Moreover, Reinhold Rudenberg, the scientific director of Siemens-Schuckertwerke, obtained the patent for the electron microscope in May 1931.

In 1932, Ernst Lubcke of Siemens & Halske built and obtained images from a prototype electron microscope, applying concepts described in the Rudenberg patent applications. Five years later (1937), the firm financed the work of Ernst Ruska and Bodo von Borries, and employed Helmut Ruska (Ernst's brother) to develop applications for the microscope, especially with biological specimens. Also in 1937, Manfred von Ardenne pioneered the scanning electron microscope. The first *practical* electron microscope was constructed in 1938, at the University of Toronto, by Eli Franklin Burton and students Cecil Hall, James Hillier, and Albert Prebus; and Siemens produced the first *commercial* transmission electron microscope (TEM) in 1939. Although contemporary electron microscopes are capable of two million-power magnification, as scientific instruments, they remain based upon Ruska's prototype.

Sample Preparation

Materials to be viewed under an electron microscope may require processing to produce a suitable sample. The technique required varies depending on the specimen and the analysis required:

- *Chemical fixation* – for biological specimens aims to stabilize the specimen's mobile macromolecular structure by chemical crosslinking of proteins with aldehydes such as formaldehyde and glutaraldehyde, and lipids with osmium tetroxide.
- *Negative stain* – suspensions containing nanoparticles or fine biological material (such as viruses and bacteria) are briefly mixed with a dilute solution of an electron-opaque solution such as ammonium molybdate, uranyl acetate (or formate), or phosphotungstic acid. This mixture is applied to a suitably coated EM grid, blotted, then allowed to dry. Viewing of this preparation in the TEM should be carried out without delay for best results. The method is important in microbiology for fast but crude morphological identification, but can also be used as the basis for high resolution 3D reconstruction using EM tomography methodology when carbon films are used for support. Negative staining is also used for observation of nanoparticles.
- *Cryofixation* – freezing a specimen so rapidly, in liquid ethane, and maintained at liquid nitrogen or even liquid helium temperatures, so that the water forms vitreous (non-crystalline) ice. This preserves the specimen in a snapshot of its solution state. An entire field called cryo-electron microscopy has branched from this technique. With the development of cryo-electron microscopy of vitreous sections (CEMOVIS), it is now possible to observe samples from virtually any biological specimen close to its native state.
- *Dehydration* – freeze drying, or replacement of water with organic solvents such as ethanol or acetone, followed by critical point drying or infiltration with embedding resins.
- *Embedding, biological specimens* – after dehydration, tissue for observation in the transmission electron microscope is embedded so it can be sectioned ready for viewing. To do this the tissue is passed through a 'transition solvent' such as Propylene oxide (epoxypropane) and then infiltrated with an epoxy resin such as Araldite, Epon, or Durcupan; tissues may also be embedded directly in water-miscible acrylic resin. After the resin has been polymerized (hardened) the sample is thin sectioned (ultrathin sections) and stained – it is then ready for viewing.
- *Embedding, materials* – after embedding in resin, the specimen is usually ground and polished to a mirror-like finish using ultra-fine

abrasives. The polishing process must be performed carefully to minimize scratches and other polishing artifacts that reduce image quality.

- *Sectioning* – produces thin slices of specimen, semitransparent to electrons. These can be cut on an ultramicrotome with a diamond knife to produce ultra-thin slices about 60–90 nm thick. Disposable glass knives are also used because they can be made in the lab and are much cheaper.
- *Staining* – uses heavy metals such as lead, uranium or tungsten to scatter imaging electrons and thus give contrast between different structures, since many (especially biological) materials are nearly "transparent" to electrons (weak phase objects). In biology, specimens can be stained "en bloc" before embedding and also later after sectioning. Typically thin sections are stained for several minutes with an aqueous or alcoholic solution of uranyl acetate followed by aqueous lead citrate.
- *Freeze-fracture or freeze-etch* – a preparation method particularly useful for examining lipid membranes and their incorporated proteins in "face on" view. The fresh tissue or cell suspension is frozen rapidly (cryofixation), then fractured by simply breaking or by using a microtome while maintained at liquid nitrogen temperature. The cold fractured surface (sometimes "etched" by increasing the temperature to about "100 °C for several minutes to let some ice sublime) is then shadowed with evaporated platinum or gold at an average angle of 45° in a high vacuum evaporator. A second coat of carbon, evaporated perpendicular to the average surface plane is often performed to improve stability of the replica coating. The specimen is returned to room temperature and pressure, then the extremely fragile "pre-shadowed" metal replica of the fracture surface is released from the underlying biological material by careful chemical digestion with acids, hypochlorite solution or SDS detergent. The still-floating replica is thoroughly washed free from residual chemicals, carefully fished up on fine grids, dried then viewed in the TEM.
- *Ion beam milling* – thins samples until they are transparent to electrons by firing ions (typically argon) at the surface from an angle and sputtering material from the surface. A subclass of this is focused ion beam milling, where gallium ions are used to produce an electron transparent membrane in a specific region of the sample, for example through a device within a microprocessor. Ion beam milling may also be used for cross-section polishing prior to SEM analysis of materials that are difficult to prepare using mechanical polishing.
- *Conductive coating* – an ultrathin coating of electrically conducting material, deposited either by high vacuum evaporation or by low vacuum sputter coating of the sample. This is done to prevent the accumulation

of static electric fields at the specimen due to the electron irradiation required during imaging. The coating materials include gold, gold/palladium, platinum, tungsten, graphite, etc.

- *Earthing* – to avoid electrical charge accumulation on a conductive coated sample, it is usually electrically connected to the metal sample holder. Often an electrically conductive adhesive is used for this purpose.

Disadvantages

Electron microscopes are expensive to build and maintain, but the capital and running costs of confocal light microscope systems now overlaps with those of basic electron microscopes. Microscopes designed to achieve high resolutions must be housed in stable buildings (sometimes underground) with special services such as magnetic field cancelling systems.

The samples largely have to be viewed in vacuum, as the molecules that make up air would scatter the electrons. One exception is the environmental scanning electron microscope, which allows hydrated samples to be viewed in a low-pressure (up to 20 Torr or 2.7 kPa) and/or wet environment.

Scanning electron microscopes operating in conventional high-vacuum mode usually image conductive specimens; therefore non-conductive materials require conductive coating (gold/palladium alloy, carbon, osmium, etc.) Low-voltage mode of modern microscopes makes possible observation of non-conductive specimens without coating. Non-conductive materials can be imaged also by a variable pressure (or environmental) scanning electron microscope.

Small, stable specimens such as carbon nanotubes, diatom frustules and small mineral crystals (asbestos fibres, for example) require no special treatment before being examined in the electron microscope. Samples of hydrated materials, including almost all biological specimens have to be prepared in various ways to stabilize them, reduce their thickness (ultrathin sectioning) and increase their electron optical contrast (staining). These processes may result in *artifacts*, but these can usually be identified by comparing the results obtained by using radically different specimen preparation methods. It is generally believed by scientists working in the field that as results from various preparation techniques have been compared and that there is no reason that they should all produce similar artifacts, it is reasonable to believe that electron microscopy features correspond with those of living cells. Since the 1980s, analysis of cryofixed, vitrified specimens has also become increasingly used by scientists, further confirming the validity of this technique.

Scanning Electron Microscope (SEM)

A scanning electron microscope (SEM) is a type of electron microscope that produces images of a sample by scanning it with a focused beam of electrons.

The electrons interact with atoms in the sample, producing various signals that can be detected and that contain information about the sample's surface topography and composition. The electron beam is generally scanned in a raster scan pattern, and the beam's position is combined with the detected signal to produce an image. SEM can achieve resolution better than 1 nanometre. Specimens can be observed in high vacuum, in low vacuum, and (in environmental SEM) in wet conditions.

The most common mode of detection is by secondary electrons emitted by atoms excited by the electron beam. The number of secondary electrons is a function of the angle between the surface and the beam. On a flat surface, the plume of secondary electrons is mostly contained by the sample, but on a tilted surface, the plume is partially exposed and more electrons are emitted. By scanning the sample and detecting the secondary electrons, an image displaying the tilt of the surface is created.

An account of the early history of SEM has been presented by McMullan. Although Max Knoll produced a photo with a 50 mm object-field-width showing channeling contrast by the use of an electron beam scanner, it was Manfred von Ardenne who in 1937 invented a true microscope with high magnification by scanning a very small raster with a demagnified and finely focused electron beam. Ardenne applied the scanning principle not only to achieve magnification but also to purposefully eliminate the chromatic aberration otherwise inherent in the electron microscope. He further discussed the various detection modes, possibilities and theory of SEM, together with the construction of the first high magnification SEM. Further work was reported by Zworykin's group, followed by the Cambridge groups in the 1950s and early 1960s headed by Charles Oatley, all of which finally led to the marketing of the first commercial instrument by Cambridge Scientific Instrument Company as the "Stereoscan" in 1965 (delivered to DuPont).

Principles and Capacities

The types of signals produced by a SEM include secondary electrons (SE), back-scattered electrons (BSE), characteristic X-rays, light (cathodoluminescence) (CL), specimen current and transmitted electrons. Secondary electron detectors are standard equipment in all SEMs, but it is rare that a single machine would have detectors for all possible signals. The signals result from interactions of the electron beam with atoms at or near the surface of the sample. In the most common or standard detection mode, secondary electron imaging or SEI, the SEM can produce very high-resolution images of a sample surface, revealing details less than 1 nm in size. Due to the very narrow electron beam, SEM micrographs have a large depth of field yielding a characteristic three-dimensional appearance useful for understanding the surface structure of a sample. This is exemplified by the micrograph of pollen shown above. A wide range of magnifications is possible, from about 10 times (about equivalent to that of a powerful hand-lens) to more than 500,000 times, about 250 times the magnification limit of the best light microscopes.

Back-scattered electrons (BSE) are beam electrons that are reflected from the sample by elastic scattering. BSE are often used in analytical SEM along with the spectra made from the characteristic X-rays, because the intensity of the BSE signal is strongly related to the atomic number (Z) of the specimen. BSE images can provide information about the distribution of different elements in the sample. For the same reason, BSE imaging can image colloidal gold immuno-labels of 5 or 10 nm diameter, which would otherwise be difficult or impossible to detect in secondary electron images in biological specimens. Characteristic X-rays are emitted when the electron beam removes an inner shell electron from the sample, causing a higher-energy electron to fill the shell and release energy. These characteristic X-rays are used to identify the composition and measure the abundance of elements in the sample.

Sample Preparation

All samples must also be of an appropriate size to fit in the specimen chamber and are generally mounted rigidly on a specimen holder called a specimen stub. Several models of SEM can examine any part of a 6-inch (15 cm) semiconductor wafer, and some can tilt an object of that size to 45°.

For conventional imaging in the SEM, specimens must be electrically conductive, at least at the surface, and electrically grounded to prevent the accumulation of electrostatic charge at the surface. Metal objects require little special preparation for SEM except for cleaning and mounting on a specimen stub. Nonconductive specimens tend to charge when scanned by the electron beam, and especially in secondary electron imaging mode, this causes scanning faults and other image artifacts. They are therefore usually coated with an ultrathin coating of electrically conducting material, deposited on the sample either by low-vacuum sputter coating or by high-vacuum evaporation. Conductive materials in current use for specimen coating include gold, gold/palladium alloy, platinum, osmium, iridium, tungsten, chromium, and graphite. Additionally, coating may increase signal/noise ratio for samples of low atomic number (Z). The improvement arises because secondary electron emission for high-Z materials is enhanced.

An alternative to coating for some biological samples is to increase the bulk conductivity of the material by impregnation with osmium using variants of the OTO staining method (O-osmium, T-thiocarbohydrazide, O-osmium).

Nonconducting specimens may be imaged uncoated using environmental SEM (ESEM) or low-voltage mode of SEM operation. Environmental SEM instruments place the specimen in a relatively high-pressure chamber where the working distance is short and the electron optical column is differentially pumped to keep vacuum adequately low at the electron gun. The high-pressure region around the sample in the ESEM neutralizes charge and provides an amplification of the secondary electron signal. Low-voltage SEM is typically conducted in an FEG-SEM because the field emission guns (FEG) is capable of producing high primary electron brightness and small spot size even at low accelerating potentials. Operating conditions to prevent charging of non-conductive specimens must be adjusted such that the incoming beam current was equal to sum of outcoming secondary and backscattered electrons currents. It usually occurs at accelerating voltages of 0.3–4 kV.

Embedding in a resin with further polishing to a mirror-like finish can be used for both biological and materials specimens when imaging in backscattered electrons or when doing quantitative X-ray microanalysis.

The main preparation techniques are not required in the environmental SEM outlined below, but some biological specimens can benefit from fixation.

Biological Samples

For SEM, a specimen is normally required to be completely dry, since the specimen chamber is at high vacuum. Hard, dry materials such as wood, bone, feathers, dried insects, or shells can be examined with little further treatment, but living cells and tissues and whole, soft-bodied organisms usually require chemical fixation to preserve and stabilize their structure. Fixation is usually performed by incubation in a solution of a buffered chemical fixative, such as glutaraldehyde, sometimes in combination with formaldehyde and other fixatives, and optionally followed by postfixation with osmium tetroxide. The fixed tissue is then dehydrated. Because air-drying causes collapse and shrinkage, this is commonly achieved by replacement of water in the cells with organic solvents such as ethanol or acetone, and replacement of these solvents in turn with a transitional fluid such as liquid carbon dioxide by critical point drying. The carbon dioxide is finally removed while in a supercritical state, so that no gas-liquid interface is present within the sample during drying. The dry specimen is usually mounted on a specimen stub using an adhesive such as epoxy resin or electrically conductive double-sided adhesive tape, and sputter-coated with gold or gold/palladium alloy before examination in the microscope.

If the SEM is equipped with a cold stage for cryo microscopy, cryofixation may be used and low-temperature scanning electron microscopy performed on the cryogenically fixed specimens. Cryo-fixed specimens may be cryo-fractured under vacuum in a special apparatus to reveal internal structure, sputter-coated, and transferred onto the SEM cryo-stage while still frozen. Low-temperature scanning electron microscopy is also applicable to the imaging of temperature-sensitive materials such as ice and fats.

Freeze-fracturing, freeze-etch or freeze-and-break is a preparation method particularly useful for examining lipid membranes and their incorporated proteins in "face on" view. The preparation method reveals the proteins embedded in the lipid bilayer.

Materials

Back scattered electron imaging, quantitative X-ray analysis, and X-ray mapping of specimens often requires that the surfaces be ground and polished to an ultra smooth surface. Specimens that undergo WDS or EDS analysis are often carbon coated. In general, metals are not coated prior to imaging in the SEM because they are conductive and provide their own pathway to ground. Fractography is the study of fractured surfaces that can be done on a light microscope or commonly, on an SEM. The fractured surface is cut to a suitable size, cleaned of any organic residues, and mounted on a specimen holder for viewing in the SEM.

Integrated circuits may be cut with a focused ion beam (FIB) or other ion beam milling instrument for viewing in the SEM. The SEM in the first case may be incorporated into the FIB.

Metals, geological specimens, and integrated circuits all may also be chemically polished for viewing in the SEM.

Special high-resolution coating techniques are required for high-magnification imaging of inorganic thin films.

Scanning Process and Image Formation

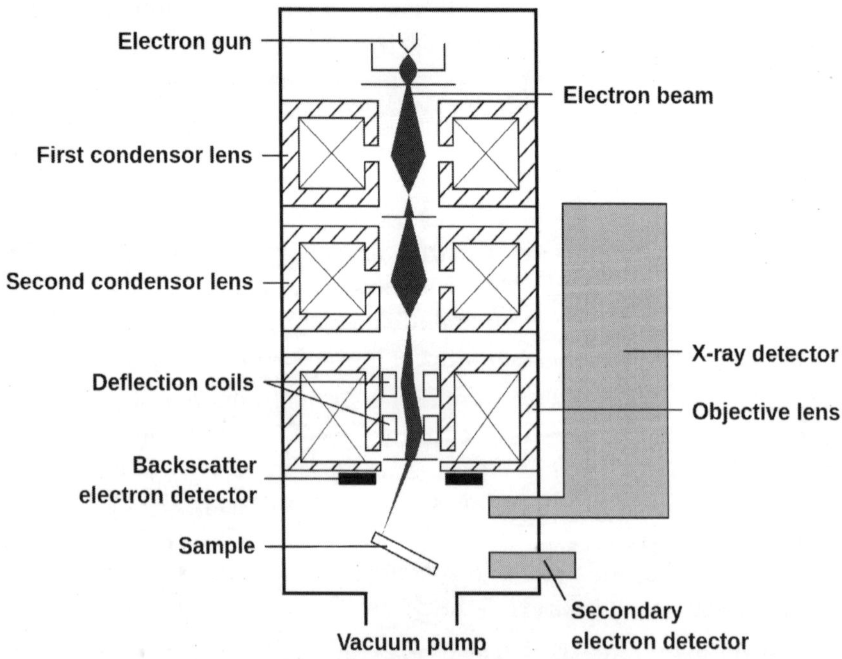

Figure: *Schematic of an SEM.*

In a typical SEM, an electron beam is thermionically emitted from an electron gun fitted with a tungsten filament cathode. Tungsten is normally used in thermionic electron guns because it has the highest melting point and lowest vapour pressure of all metals, thereby allowing it to be heated for electron emission, and because of its low cost. Other types of electron emitters include lanthanum hexaboride.

The electron beam, which typically has an energy ranging from 0.2 keV to 40 keV, is focused by one or two condenser lenses to a spot about 0.4 nm to 5 nm in diameter. The beam passes through pairs of scanning coils or pairs of deflector plates in the electron column, typically in the final lens, which deflect the beam in the x and y axes so that it scans in a raster fashion over a rectangular area of the sample surface.

When the primary electron beam interacts with the sample, the electrons lose energy by repeated random scattering and absorption within a teardrop-shaped volume of the specimen known as the interaction volume, which extends from less than 100 nm to approximately 5 μm into the surface. The size of the interaction volume depends on the electron's landing energy, the atomic number of the specimen and the specimen's density. The energy exchange between the electron beam and the sample results in the reflection of high-energy electrons by elastic scattering, emission of secondary electrons by inelastic scattering and the emission of electromagnetic radiation, each of which can be detected by specialised detectors. The beam current absorbed by the specimen can also be detected and used to create images of the distribution of specimen current. Electronic amplifiers of various types are used to amplify the signals, which are displayed as variations in brightness on a computer monitor (or, for vintage models, on a cathode ray tube). Each pixel of computer videomemory is synchronized with the position of the beam on the specimen in the microscope, and the resulting image is therefore a distribution map of the intensity of the signal being emitted from the scanned area of the specimen. In older microscopes image may be captured by photography from a high-resolution cathode ray tube, but in modern machines image is saved to a computer data storage.

Magnification

Magnification in a SEM can be controlled over a range of up to 6 orders of magnitude from about 10 to 500,000 times. Unlike optical and transmission electron microscopes, image magnification in the SEM is not a function of the power of the objective lens. SEMs may have condenser and objective lenses, but their function is to focus the beam to a spot, and not to image the specimen. Provided the electron gun can generate a beam with sufficiently small diameter, a SEM could in principle work entirely without condenser or objective lenses, although it might not be very versatile or achieve very high resolution. In a SEM, as in scanning probe microscopy, magnification results from the ratio of the dimensions of the raster on the specimen and the raster on the display device. Assuming that the display screen has a fixed size, higher magnification results from reducing the size of the raster on the specimen, and vice versa. Magnification is therefore controlled by the current supplied to the x, y scanning coils, or the voltage supplied to the x, y deflector plates, and not by objective lens power.

Colour

The most common configuration for an SEM produces a single value per pixel, with the results usually rendered as black-and-white images. However, often these images are then colourized through the use of feature-detection software, or simply by hand-editing using a graphics editor. This is usually for aesthetic effect, for clarifying structure, or for adding a realistic appearance to the sample and generally does not add information about the specimen.

In some configurations more information is gathered per pixel, often by the use of multiple detectors. The attributes of topography and material contrast can be obtained by a pair of backscattered electron detectors and such attributes can be superimposed on a single colour image by assigning a different primary colour to each attribute. Similarly, a combination of backscattered and secondary electron signals can be assigned to different colours and superimposed on a single colour micrograph displaying simultaneously the properties of the specimen.

In a similar method, secondary electron and backscattered electron detectors are superimposed and a colour is assigned to each of the images captured by each detector, with an end result of a combined colour image where colours are related to the density of the components. This method is known as density-dependent colour SEM (DDC-SEM). Micrographs produced by DDC-SEM retain topographical information, which is better captured by the secondary electrons detector and combine it to the information about density, obtained by the backscattered electron detector.

Some types of detectors used in SEM have analytical capabilities, and can provide several items of data at each pixel. Examples are the Energy-dispersive X-ray spectroscopy (EDS) detectors used in elemental analysis and Cathodoluminescence microscope (CL) systems that analyse the intensity and spectrum of electron-induced luminescence in (for example) geological specimens. In SEM systems using these detectors it is common to colour code the signals and superimpose them in a single colour image, so that differences in the distribution of the various components of the specimen can be seen clearly and compared. Optionally, the standard secondary electron image can be merged with the one or more compositional channels, so that the specimen's structure and composition can be compared. Such images can be made while maintaining the full integrity of the original signal, which is not modified in any way.

Detection of Secondary Electrons

The most common imaging mode collects low-energy (<50 eV) secondary electrons that are ejected from the k-shell of the specimen atoms by inelastic scattering interactions with beam electrons. Due to their low energy, these electrons originate within a few nanometres from the sample surface. The electrons are detected by an Everhart-Thornley detector, which is a type of scintillator-photomultiplier system. The secondary electrons are first collected by attracting them towards an electrically biased grid at about +400 V, and then further accelerated towards a phosphor or scintillator positively biased to about +2,000 V. The accelerated secondary electrons are now sufficiently energetic to cause the scintillator to emit flashes of light (cathodoluminescence), which are conducted to a photomultiplier outside the SEM column via a light pipe and a window in the wall of the specimen chamber. The amplified electrical signal output by the photomultiplier is displayed as a two-dimensional intensity distribution that can be viewed and photographed on an analogue video display, or subjected to analog-to-digital conversion and displayed and saved as a digital image. This process relies on a raster-scanned primary beam. The brightness of the signal depends on the number of secondary electrons reaching the detector. If the beam enters the sample perpendicular to the surface, then the activated region is uniform about the axis of the beam and a certain number of electrons "escape" from within the sample. As the angle of incidence increases, the "escape" distance of one side of the beam will decrease, and more secondary electrons will be emitted. Thus steep surfaces and edges tend to be brighter than flat surfaces, which results in images with a well-defined, three-dimensional appearance. Using the signal of secondary electrons image resolution less than 0.5 nm is possible.

Detection of Backscattered Electrons

Backscattered electrons (BSE) consist of high-energy electrons originating in the electron beam, that are reflected or back-scattered out of the specimen interaction volume by elastic scattering interactions with specimen atoms. Since heavy elements (high atomic number) backscatter electrons more strongly than light elements (low atomic number), and thus appear brighter in the image, BSE are used to detect contrast between areas with different chemical compositions. The Everhart-Thornley detector, which is normally positioned to one side of the specimen, is inefficient for the detection of backscattered electrons because few such electrons are emitted in the solid angle subtended by the detector, and because the positively biased detection grid has little ability to attract the higher energy BSE. Dedicated backscattered electron detectors are positioned above the sample in a "doughnut" type arrangement, concentric with the electron beam, maximizing the solid angle of collection. BSE detectors are usually either of scintillator or of semiconductor types. When all parts of the detector are used to collect electrons symmetrically about the beam, atomic number contrast is produced. However, strong topographic contrast is produced by collecting back-scattered electrons from one side above the specimen using an asymmetrical, directional BSE detector; the resulting contrast appears as illumination of the topography from that side. Semiconductor detectors can be made in radial segments that can be switched in or out to control the type of contrast produced and its directionality. Backscattered electrons can also be used to form an electron backscatter diffraction (EBSD) image that can be used to determine the crystallographic structure of the specimen.

Beam-Injection Analysis of Semiconductors

The nature of the SEM's probe, energetic electrons, makes it uniquely suited to examining the optical and electronic properties of semiconductor materials. The high-energy electrons from the SEM beam will inject charge carriers into the semiconductor. Thus, beam electrons lose energy by promoting electrons from the valence band into the conduction band, leaving behind holes.

In a direct bandgap material, recombination of these electron-hole pairs will result in cathodoluminescence; if the sample contains an internal electric field, such as is present at a p-n junction, the SEM beam injection of carriers will cause electron beam induced current (EBIC) to flow.

Cathodoluminescence and EBIC are referred to as "beam-injection" techniques, and are very powerful probes of the optoelectronic behaviour of semiconductors, in particular for studying nanoscale features and defects.

Cathodoluminescence

Cathodoluminescence, the emission of light when atoms excited by high-energy electrons return to their ground state, is analogous to UV-induced fluorescence, and some materials such as zinc sulfide and some fluorescent dyes, exhibit both phenomena. Over the last decades, cathodoluminescence was most commonly experienced as the light emission from the inner surface of the cathode ray tube in television sets and computer CRT monitors. In the SEM, CL detectors either collect all light emitted by the specimen or can analyse the wavelengths emitted by the specimen and display an emission spectrum or an image of the distribution of cathodoluminescence emitted by the specimen in real colour.

X-Ray Microanalysis

X-rays, which are produced by the interaction of electrons with the sample, may also be detected in an SEM equipped for energy-dispersive X-ray spectroscopy or wavelength dispersive X-ray spectroscopy.

Resolution of the SEM

The spatial resolution of the SEM depends on the size of the electron spot, which in turn depends on both the wavelength of the electrons and the electron-optical system that produces the scanning beam. The resolution is also limited by the size of the interaction volume, or the extent to which the material interacts with the electron beam. The spot size and the interaction volume are both large compared to the distances between atoms, so the resolution of the SEM is not high enough to image individual atoms, as is possible in the shorter wavelength (i.e. higher energy) transmission electron microscope (TEM). The SEM has compensating advantages, though, including the ability to image a comparatively large area of the specimen; the ability to image bulk materials (not just thin films or foils); and the variety of analytical modes available for measuring the composition and properties of the specimen. Depending on the instrument, the resolution can fall somewhere between less than 1 nm and 20 nm. By 2009, The world's highest SEM resolution at high-beam energies (0.4 nm at 30 kV) is obtained with the Hitachi SU-9000.

Environmental SEM

Conventional SEM requires samples to be imaged under vacuum, because a gas atmosphere rapidly spreads and attenuates electron beams. As a consequence, samples that produce a significant amount of vapour, e.g. wet biological samples or oil-bearing rock, must be either dried or cryogenically frozen. Processes involving phase transitions, such as the drying of adhesives or melting of alloys, liquid transport, chemical reactions, and solid-air-gas systems, in general cannot be observed. Some observations of living insects have been possible, however.

The first commercial development of the ESEM in the late 1980s allowed samples to be observed in low-pressure gaseous environments (e.g. 1–50 Torr or 0.1–6.7 kPa) and high relative humidity (up to 100%). This was made possible by the development of a secondary-electron detector capable of operating in the presence of water vapour and by the use of pressure-limiting apertures with differential pumping in the path of the electron beam to separate the vacuum region (around the gun and lenses) from the sample chamber.

The first commercial ESEMs were produced by the ElectroScan Corporation in USA in 1988. ElectroScan was taken over by Philips (who later sold their electron-optics division to FEI Company) in 1996.

ESEM is especially useful for non-metallic and biological materials because coating with carbon or gold is unnecessary. Uncoated Plastics and Elastomers can be routinely examined, as can uncoated biological samples. Coating can be difficult to reverse, may conceal small features on the surface of the sample and may reduce the value of the results obtained. X-ray analysis is difficult with a coating of a heavy metal, so carbon coatings are routinely used in conventional SEMs, but ESEM makes it possible to perform X-ray microanalysis on uncoated non-conductive specimens. ESEM may be the preferred for electron microscopy of unique samples from criminal or civil actions, where forensic analysis may need to be repeated by several different experts.

3D in SEM

SEMs do not naturally provide 3D images contrary to SPMs. However 3D data can be obtained using a SEM with different methods such as:

- photogrammetry (2 or 3 images from tilted specimen)
- photometric stereo also called "shape from shading" (use of 4 images from BSE detector)
- inverse reconstruction using electron-material interactive models
- vertical stacks of SEM micrographs plus image-processing software

Possible applications are roughness measurement, measurement of fractal dimension, corrosion measurement and step height evaluation.

Transmission Electron Microscope TEM

Transmission electron microscopy (TEM) is a microscopy technique in which a beam of electrons is transmitted through an ultra-thin specimen, interacting with the specimen as it passes through. An image is formed from the interaction of the electrons transmitted through the specimen; the image is magnified and focused onto an imaging device, such as a fluorescent screen, on a layer of photographic film, or to be detected by a sensor such as a CCD camera.

TEMs are capable of imaging at a significantly higher resolution than light microscopes, owing to the small de Broglie wavelength of electrons. This enables the instrument's user to examine fine detail—even as small as a single column of atoms, which is thousands of times smaller than the smallest resolvable object in a light microscope. TEM forms a major analysis method in a range of scientific fields, in both physical and biological sciences. TEMs find application in cancer research, virology, materials science as well as pollution, nanotechnology, and semiconductor research.

At smaller magnifications TEM image contrast is due to absorption of electrons in the material, due to the thickness and composition of the material. At higher magnifications complex wave interactions modulate the intensity of the image, requiring expert analysis of observed images. Alternate modes of use allow for the TEM to observe modulations in chemical identity, crystal orientation, electronic structure and sample induced electron phase shift as well as the regular absorption based imaging.

The first TEM was built by Max Knoll and Ernst Ruska in 1931, with this group developing the first TEM with resolution greater than that of light in 1933 and the first commercial TEM in 1939.

Initial Development

Ernst Abbe originally proposed that the ability to resolve detail in an object was limited approximately by the wavelength of the light used in imaging, which limits the resolution of an optical microscope to a few hundred nanometres. Developments into ultraviolet (UV) microscopes, led by Köhler and Rohr, allowed for an increase in resolving power of about a factor of two. However this required more expensive quartz optical components, due to the absorption of UV by glass. At this point it was believed that obtaining an image with sub-micrometre information was simply impossible due to this wavelength constraint.

It had earlier been recognised by Plücker in 1858 that the deflection of "cathode rays" (electrons) was possible by the use of magnetic fields. This effect had been utilised to build primitive cathode ray oscilloscopes (CROs) as early as 1897 by Ferdinand Braun, intended as a measurement device. Indeed in 1891 it was recognised by Riecke that the cathode rays could be focused by these magnetic fields, allowing for simple lens designs. Later this theory was extended by Hans Busch in his work published in 1926, who showed that the lens maker's equation, could under appropriate assumptions, be applicable to electrons.

In 1928, at the Technological University of Berlin Adolf Matthias, Professor of High voltage Technology and Electrical Installations, appointed Max Knoll to lead a team of researchers to advance the CRO design. The team consisted of several PhD students including Ernst Ruska and Bodo von Borries. This team of researchers concerned themselves with lens design and CRO column placement, which they attempted to obtain the parameters that could be optimised to allow for construction of better CROs, as well as the development of electron optical components which could be used to generate low magnification (nearly 1:1) images. In 1931 the group successfully generated magnified images of mesh grids placed over the anode aperture. The device used two magnetic lenses to achieve higher magnifications, arguably the first electron microscope. In that same year, Reinhold Rudenberg, the scientific director of the Siemens company, had patented an electrostatic lens electron microscope.

Improving Resolution

At this time the wave nature of electrons, which were considered charged matter particles, had not been fully realised until the publication of the De Broglie hypothesis in 1927. The group was unaware of this publication until 1932, where it was quickly realised that the De Broglie wavelength of electrons was many orders of magnitude smaller than that for light, theoretically allowing for imaging at atomic scales. In April 1932, Ruska suggested the construction of a new electron microscope for direct imaging of specimens inserted into the microscope, rather than simple mesh grids or images of apertures. With this device successful diffraction and normal imaging of aluminium sheet was achieved, however exceeding the magnification achievable with light microscopy had still not been successfully demonstrated. This goal was achieved in September 1933, using images of cotton fibers, which were quickly acquired before being damaged by the electron beam.

At this time, interest in the electron microscope had increased, with other groups, such as Paul Anderson and Kenneth Fitzsimmons of Washington State University, and Albert Prebus and James Hillier at the University of Toronto who constructed the first TEMs in North America in 1935 and 1938, respectively, continually advancing TEM design.

Research continued on the electron microscope at Siemens in 1936, the aim of the research was the development improvement of TEM imaging properties, particularly with regard to biological specimens. At this time electron microscopes were being fabricated for specific groups, such as the "EM1" device used at the UK National Physical Laboratory. In 1939 the first commercial electron microscope, pictured, was installed in the Physics department of I. G Farben-Werke. Further work on the electron microscope was hampered by the destruction of a new laboratory constructed at Siemens by an air-raid, as well as the death of two of the researchers, Heinz Müller and Friedrick Krause during World War II.

Further Research

After World War II, Ruska resumed work at Siemens, where he continued to develop the electron microscope, producing the first microscope with 100k magnification. The fundamental structure of this microscope design, with multi-stage beam preparation optics, is still used in modern microscopes. The worldwide electron microscopy community advanced with electron microscopes being manufactured in Manchester UK, the USA (RCA), Germany (Siemens) and Japan. The first international conference in electron microscopy was in Delft in 1949, with more than one hundred attendees. Later conferences included the "First" international conference in Paris, 1950 and then in London in 1954.

With the development of TEM, the associated technique of scanning transmission electron microscopy (STEM) was re-investigated and did not become developed until the 1970s, with Albert Crewe at the University of Chicago developing the field emission gun and adding a high quality objective lens to create the modern STEM. Using this design, Crewe demonstrated the ability to image atoms using annular dark-field imaging. Crewe and coworkers at the University of Chicago developed the cold field electron emission source and built a STEM able to visualize single heavy atoms on thin carbon substrates. In 2008, Jannick Meyer et al. described the direct visualization of light atoms such as carbon and even hydrogen using TEM and a clean single-layer graphene substrate.

Background

Electrons: Theoretically, the maximum resolution, *d*, that one can obtain with a light microscope has been limited by the wavelength of the photons that are being used to probe the sample, λ and the numerical aperture of the system, *NA*.

$$d = \frac{\lambda}{2n\sin\alpha} \approx \frac{\lambda}{2\mathrm{NA}}$$

Early twentieth century scientists theorised ways of getting around the limitations of the relatively large wavelength of visible light (wavelengths of 400–700 nanometres) by using electrons. Like all matter, electrons have both wave and particle properties (as theorized by Louis-Victor de Broglie), and their wave-like properties mean that a beam of electrons can be made to behave like a beam of electromagnetic radiation. The wavelength of electrons is related to their kinetic energy via the de Broglie equation. An additional correction must be made to account for relativistic effects, as in a TEM an electron's velocity approaches the speed of light, *c*.

$$\lambda_e \approx \frac{h}{\sqrt{2m_0E\left(1+\frac{E}{2m_0c^2}\right)}}$$

where, h is Planck's constant, m_0 is the rest mass of an electron and E is the energy of the accelerated electron. Electrons are usually generated in an electron microscope by a process known as thermionic emission from a filament, usually tungsten, in the same manner as a light bulb, or alternatively by field electron emission. The electrons are then accelerated by an electric potential (measured in volts) and focused by electrostatic and electromagnetic lenses onto the sample. The transmitted beam contains information about electron density, phase and periodicity; this beam is used to form an image.

Source Formation

From the top down, the TEM consists of an emission source, which may be a tungsten filament, or a lanthanum hexaboride (LaB_6) source. For tungsten, this will be of the form of either a hairpin-style filament, or a small spike-shaped filament. LaB_6 sources utilise small single crystals. By connecting this gun to a high voltage source (typically ~100–300 kV) the gun will, given sufficient current, begin to emit electrons either by thermionic or field electron emission into the vacuum. This extraction is usually aided by the use of a Wehnelt cylinder. Once extracted, the upper lenses of the TEM allow for the formation of the electron probe to the desired size and location for later interaction with the sample.

Manipulation of the electron beam is performed using two physical effects. The interaction of electrons with a magnetic field will cause electrons to move according to the left hand rule, thus allowing for electromagnets to manipulate the electron beam. The use of magnetic fields allows for the formation of a magnetic lens of variable focusing power, the lens shape originating due to the distribution of magnetic flux. Additionally, electrostatic fields can cause the electrons to be deflected through a constant angle. Coupling of two deflections in opposing directions with a small intermediate gap allows for the formation of a shift in the beam path, this being used in TEM for beam shifting, subsequently this is extremely important to STEM. From these two effects, as well as the use of an electron imaging system, sufficient control over the beam path is possible for TEM operation. The optical configuration of a TEM can be rapidly changed, unlike that for an optical microscope, as lenses in the beam path can be enabled, have their strength changed, or be disabled entirely simply via rapid electrical switching, the speed of which is limited by effects such as the magnetic hysteresis of the lenses.

Optics

The lenses of a TEM allow for beam convergence, with the angle of convergence as a variable parameter, giving the TEM the ability to change magnification simply by modifying the amount of current that flows through the coil, quadrupole or hexapole lenses. The quadrupole lens is an arrangement of electromagnetic coils at the vertices of the square, enabling the generation of a lensing magnetic fields, the hexapole configuration simply enhances the lens symmetry by using six, rather than four coils.

Typically a TEM consists of three stages of lensing. The stages are the condensor lenses, the objective lenses, and the projector lenses. The condensor lenses are responsible for primary beam formation, while the objective lenses focus the beam that comes through the sample itself (in STEM scanning mode, there are also objective lenses above the sample to make the incident electron beam convergent). The projector lenses are used to expand the beam onto the phosphor screen or other imaging device, such as film. The magnification of the TEM is due to the ratio of the distances between the specimen and the objective lens' image plane. Additional quad or hexapole lenses allow for the correction of asymmetrical beam distortions, known as astigmatism. It is noted that TEM optical configurations differ significantly with implementation, with manufacturers using custom lens configurations, such as in spherical aberration corrected instruments, or TEMs utilising energy filtering to correct electron chromatic aberration.

Display

Imaging systems in a TEM consist of a phosphor screen, which may be made of fine (10–100 μm) particulate zinc sulphide, for direct observation by the operator. Optionally, an image recording system such as film based or doped YAG screen coupled CCDs. Typically these devices can be removed or inserted into the beam path by the operator as required.

Components

A TEM is composed of several components, which include a vacuum system in which the electrons travel, an electron emission source for generation of the electron stream, a series of electromagnetic lenses, as well as electrostatic plates. The latter two allow the operator to guide and manipulate the beam as required. Also required is a device to allow the insertion into, motion within, and removal of specimens from the beam path. Imaging devices are subsequently used to create an image from the electrons that exit the system.

Vacuum System

To increase the mean free path of the electron gas interaction, a standard TEM is evacuated to low pressures, typically on the order of 10^{-4} Pa. The need for this is twofold: first the allowance for the voltage difference between the cathode and the ground without generating an arc, and secondly to reduce the collision frequency of electrons with gas atoms to negligible levels—this effect is characterised by the mean free path. TEM components such as specimen holders and film cartridges must be routinely inserted or replaced requiring a system with the ability to re-evacuate on a regular basis. As such, TEMs are equipped with multiple pumping systems and airlocks and are not permanently vacuum sealed.

The vacuum system for evacuating a TEM to an operating pressure level consists of several stages. Initially a low or roughing vacuum is achieved with either a rotary vane pump or diaphragm pumps bringing the TEM to a sufficiently low pressure to allow the operation of a turbomolecular or diffusion pump which brings the TEM to its high vacuum level necessary for operations. To allow for the low vacuum pump to not require continuous operation, while continually operating the turbomolecular pumps, the vacuum side of a low-pressure pump may be connected to chambers which accommodate the exhaust gases from the turbomolecular pump. Sections of the TEM may be isolated by the use of pressure-limiting apertures, to allow for different vacuum levels in specific areas, such as a higher vacuum of 10^{-4} to 10^{-7} Pa or higher in the electron gun in high-resolution or field-emission TEMs.

High-voltage TEMs require ultra-high vacuums on the range of 10^{-7} to 10^{-9} Pa to prevent generation of an electrical arc, particularly at the TEM cathode. As such for higher voltage TEMs a third vacuum system may operate, with the gun isolated from the main chamber either by use of gate valves or by the use of a differential pumping aperture. The differential pumping aperture is a small hole that prevents diffusion of gas molecules into the higher vacuum gun area faster than they can be pumped out. For these very low pressures either an ion pump or a getter material is used.

Poor vacuum in a TEM can cause several problems, from deposition of gas inside the TEM onto the specimen as it is being viewed through a process known as electron beam induced deposition, or in more severe cases damage to the cathode from an electrical discharge. Vacuum problems due to specimen sublimation are limited by the use of a cold trap to adsorb sublimated gases in the vicinity of the specimen.

Specimen Stage

TEM specimen stage designs include airlocks to allow for insertion of the specimen holder into the vacuum with minimal increase in pressure in other areas of the microscope. The specimen holders are adapted to hold a standard size of grid upon which the sample is placed or a standard size of self-supporting specimen. Standard TEM grid sizes are a 3.05 mm diameter ring, with a thickness and mesh size ranging from a few to 100 μm. The sample is placed onto the inner meshed area having diameter of approximately 2.5 mm. Usual grid materials are copper, molybdenum, gold or platinum. This grid is placed into the sample holder, which is paired with the specimen stage. A wide variety of designs of stages and holders exist, depending upon the type of experiment being performed. In addition to 3.05 mm grids, 2.3 mm grids are sometimes, if rarely, used. These grids were particularly used in the mineral sciences where a large degree of tilt can be required and where specimen material may be extremely rare. Electron transparent specimens have a thickness around 100 nm, but this value depends on the accelerating voltage.

Once inserted into a TEM, the sample often has to be manipulated to present the region of interest to the beam, such as in single grain diffraction, in a specific orientation. To accommodate this, the TEM stage includes mechanisms for the translation of the sample in the XY plane of the sample, for Z height adjustment of the sample holder, and usually for at least one rotation degree of freedom for the sample. Thus a TEM stage may provide four degrees of freedom for the motion of the specimen. Most modern TEMs provide the ability for two orthogonal rotation angles of movement with specialised holder designs called double-tilt sample holders. Of note however is that some stage designs, such as top-entry or vertical insertion stages once common for high resolution TEM studies, may simply only have X-Y translation available.

The design criteria of TEM stages are complex, owing to the simultaneous requirements of mechanical and electron-optical constraints and have thus generated many unique implementations.

A TEM stage is required to have the ability to hold a specimen and be manipulated to bring the region of interest into the path of the electron beam. As the TEM can operate over a wide range of magnifications, the stage must simultaneously be highly resistant to mechanical drift, with drift requirements as low as a few nm/minute while being able to move several μm/minute, with repositioning accuracy on the order of nanometres. Earlier designs of TEM accomplished this with a complex set of mechanical downgearing devices, allowing the operator to finely control the motion of the stage by several rotating rods. Modern devices may use electrical stage designs, using screw gearing in concert with stepper motors, providing the operator with a computer-based stage input, such as a joystick or trackball.

Two main designs for stages in a TEM exist, the side-entry and top entry version. Each design must accommodate the matching holder to allow for specimen insertion without either damaging delicate TEM optics or allowing gas into TEM systems under vacuum.

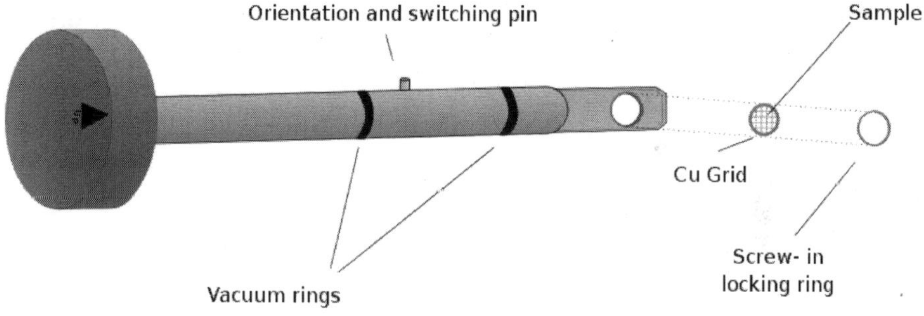

Figure: *A diagram of a single axis tilt sample holder for insertion into a TEM goniometre. Titling of the holder is achieved by rotation of the entire goniometre*

The most common is the side entry holder, where the specimen is placed near the tip of a long metal (brass or stainless steel) rod, with the specimen placed flat in a small bore. Along the rod are several polymer vacuum rings to allow for the formation of a vacuum seal of sufficient quality, when inserted into the stage. The stage is thus designed to accommodate the rod, placing the sample either in between or near the objective lens, dependent upon the objective design. When inserted into the stage, the side entry holder has its tip contained within the TEM vacuum, and the base is presented to atmosphere, the airlock formed by the vacuum rings.

Insertion procedures for side-entry TEM holders typically involve the rotation of the sample to trigger micro switches that initiate evacuation of the airlock before the sample is inserted into the TEM column.

The second design is the top-entry holder consists of a cartridge that is several cm long with a bore drilled down the cartridge axis. The specimen is loaded into the bore, possibly utilising a small screw ring to hold the sample in place. This cartridge is inserted into an airlock with the bore perpendicular to the TEM optic axis. When sealed, the airlock is manipulated to push the cartridge such that the cartridge falls into place, where the bore hole becomes aligned with the beam axis, such that the beam travels down the cartridge bore and into the specimen. Such designs are typically unable to be tilted without blocking the beam path or interfering with the objective lens.

Electron Gun

The electron gun is formed from several components: the filament, a biasing circuit, a Wehnelt cap, and an extraction anode. By connecting the filament to the negative component power supply, electrons can be "pumped" from the electron gun to the anode plate, and TEM column, thus completing the circuit. The gun is designed to create a beam of electrons exiting from the assembly at some given angle, known as the gun divergence semiangle, α. By constructing the Wehnelt cylinder such that it has a higher negative charge than the filament itself, electrons that exit the filament in a diverging manner are, under proper operation, forced into a converging pattern the minimum size of which is the gun crossover diameter.

The thermionic emission current density, J, can be related to the work function of the emitting material via Richardson's law

$$J = AT^2 \exp\left(-\frac{\Phi}{kT}\right),$$

where A is a constant, Φ is the work function and T is the temperature of the material.

This equation shows that in order to achieve sufficient current density it is necessary to heat the emitter, taking care not to cause damage by application of excessive heat, for this reason materials with either a high melting point, such as tungsten, or those with a low work function (LaB_6) are required for the gun filament. Furthermore both lanthanum hexaboride and tungsten thermionic sources must be heated in order to achieve thermionic emission, this can be achieved by the use of a small resistive strip. To prevent thermal shock, there is often a delay enforced in the application of current to the tip, to prevent thermal gradients from damaging the filament, the delay is usually a few seconds for LaB_6, and significantly lower for tungsten.

Electron Lens

Electron lenses are designed to act in a manner emulating that of an optical lens, by focusing parallel rays at some constant focal length. Lenses may operate electrostatically or magnetically. The majority of electron lenses for TEM utilise electromagnetic coils to generate a convex lens. For these lenses the field produced for the lens must be radially symmetric, as deviation from the radial symmetry of the magnetic lens causes aberrations such as astigmatism, and worsens spherical and chromatic aberration. Electron lenses are manufactured from iron, iron-cobalt or nickel cobalt alloys, such as permalloy. These are selected for their magnetic properties, such as magnetic saturation, hysteresis and permeability.

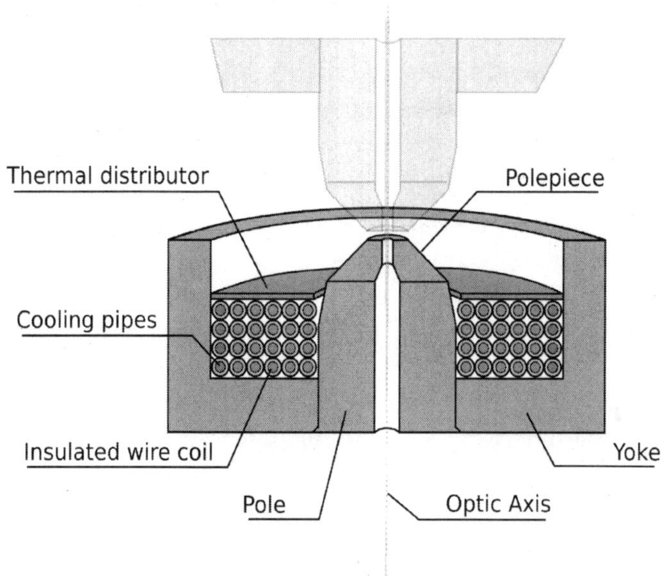

Figure: *Diagram of a TEM split polepiece design lens*

The components include the yoke, the magnetic coil, the poles, the polepiece, and the external control circuitry. The polepiece must be manufactured in a very symmetrical manner, as this provides the boundary conditions for the magnetic field that forms the lens. Imperfections in the manufacture of the polepiece can induce severe distortions in the magnetic field symmetry, which induce distortions that will ultimately limit the lenses' ability to reproduce the object plane. The exact dimensions of the gap, pole piece internal diameter and taper, as well as the overall design of the lens is often performed by finite element analysis of the magnetic field, whilst considering the thermal and electrical constraints of the design.

The coils which produce the magnetic field are located within the lens yoke. The coils can contain a variable current, but typically utilise high voltages, and therefore require significant insulation in order to prevent short-circuiting the lens components. Thermal distributors are placed to ensure the extraction of the heat generated by the energy lost to resistance of the coil windings. The windings may be water-cooled, using a chilled water supply in order to facilitate the removal of the high thermal duty.

Apertures

Apertures are annular metallic plates, through which electrons that are further than a fixed distance from the optic axis may be excluded. These consist of a small metallic disc that is sufficiently thick to prevent electrons from passing through the disc, whilst permitting axial electrons. This permission of central electrons in a TEM causes two effects simultaneously: firstly, apertures decrease the beam intensity as electrons are filtered from the beam, which may be desired in the case of beam sensitive samples. Secondly, this filtering removes electrons that are scattered to high angles, which may be due to unwanted processes such as spherical or chromatic aberration, or due to diffraction from interaction within the sample.

Apertures are either a fixed aperture within the column, such as at the condensor lens, or are a movable aperture, which can be inserted or withdrawn from the beam path, or moved in the plane perpendicular to the beam path. Aperture assemblies are mechanical devices which allow for the selection of different aperture sizes, which may be used by the operator to trade off intensity and the filtering effect of the aperture. Aperture assemblies are often equipped with micrometres to move the aperture, required during optical calibration.

Imaging Methods

Imaging methods in TEM utilise the information contained in the electron waves exiting from the sample to form an image. The projector lenses allow for the correct positioning of this electron wave distribution onto the viewing system. The observed intensity of the image, I, assuming sufficiently high quality of imaging device, can be approximated as proportional to the time-average amplitude of the electron wavefunctions, where the wave which form the exit beam is denoted by Ψ.

$$I(x) = \frac{k}{t_1 - t_0} \int_{t_0}^{t_1} \Psi\Psi^* \, dt$$

Different imaging methods therefore attempt to modify the electron waves exiting the sample in a form that is useful to obtain information with regards to the sample, or beam itself. From the previous equation, it can be deduced that the observed image depends not only on the amplitude of beam, but also on the phase of the electrons, although phase effects may often be ignored at lower magnifications. Higher resolution imaging requires thinner samples and higher energies of incident electrons. Therefore the sample can no longer be considered to be absorbing electrons, via a Beer's law effect, rather the sample can be modelled as an object that does not change the amplitude of the incoming electron wavefunction. Rather the sample modifies the phase of the incoming wave; this model is known as a pure phase object, for sufficiently thin specimens phase effects dominate the image, complicating analysis of the observed intensities. For example, to improve the contrast in the image the TEM may be operated at a slight defocus to enhance contrast, owing to convolution by the contrast transfer function of the TEM, which would normally decrease contrast if the sample was not a weak phase object.

Contrast Formation

Contrast formation in the TEM depends greatly on the mode of operation. Complex imaging techniques, which utilise the unique ability to change lens strength or to deactivate a lens, allow for many operating modes. These modes may be used to discern information that is of particular interest to the investigator.

Bright Field

The most common mode of operation for a TEM is the bright field imaging mode. In this mode the contrast formation, when considered classically, is formed directly by occlusion and absorption of electrons in the sample. Thicker regions of the sample, or regions with a higher atomic number will appear dark, whilst regions with no sample in the beam path will appear bright – hence the term “bright field”. The image is in effect assumed to be a simple two dimensional projection of the sample down the optic axis, and to a first approximation may be modelled via Beer’s law, more complex analyses require the modelling of the sample to include phase information.

Diffraction Contrast

Samples can exhibit diffraction contrast, whereby the electron beam undergoes Bragg scattering, which in the case of a crystalline sample, disperses electrons into discrete locations in the back focal plane. By the placement of apertures in the back focal plane, i.e. the objective aperture, the desired Bragg reflections can be selected (or excluded), thus only parts of the sample that are causing the electrons to scatter to the selected reflections will end up projected onto the imaging apparatus.

If the reflections that are selected do not include the unscattered beam (which will appear up at the focal point of the lens), then the image will appear dark wherever no sample scattering to the selected peak is present, as such a region without a specimen will appear dark. This is known as a dark-field image.

Modern TEMs are often equipped with specimen holders that allow the user to tilt the specimen to a range of angles in order to obtain specific diffraction conditions, and apertures placed above the specimen allow the user to select electrons that would otherwise be diffracted in a particular direction from entering the specimen. Applications for this method include the identification of lattice defects in crystals. By carefully selecting the orientation of the sample, it is possible not just to determine the position of defects but also to determine the type of defect present.

If the sample is oriented so that one particular plane is only slightly tilted away from the strongest diffracting angle (known as the Bragg Angle), any distortion of the crystal plane that locally tilts the plane to the Bragg angle will produce particularly strong contrast variations. However, defects that produce only displacement of atoms that do not tilt the crystal to the Bragg angle (i. e. displacements parallel to the crystal plane) will not produce strong contrast.

Electron Energy Loss

Utilising the advanced technique of EELS, for TEMs appropriately equipped electrons can be rejected based upon their voltage (which, due to constant charge is their energy), using magnetic sector based devices known as EELS spectrometres. These devices allow for the selection of particular energy values, which can be associated with the way the electron has interacted with the sample. For example different elements in a sample result in different electron energies in the beam after the sample. This normally results in chromatic aberration – however this effect can, for example, be used to generate an image which provides information on elemental composition, based upon the atomic transition during electron-electron interaction.

EELS spectrometres can often be operated in both spectroscopic and imaging modes, allowing for isolation or rejection of elastically scattered beams. As for many images inelastic scattering will include information that may not be of interest to the investigator thus reducing observable signals of interest, EELS imaging can be used to enhance contrast in observed images, including both bright field and diffraction, by rejecting unwanted components.

Phase Contrast

Crystal structure can also be investigated by high-resolution transmission electron microscopy (HRTEM), also known as phase contrast. When utilising a Field emission source and a specimen of uniform thickness, the images are formed due to differences in phase of electron waves, which is caused by specimen interaction. Image formation is given by the complex modulus of the incoming electron beams. As such, the image is not only dependent on the number of electrons hitting the screen, making direct interpretation of phase contrast images more complex. However this effect can be used to an advantage, as it can be manipulated to provide more information about the sample, such as in complex phase retrieval techniques.

Diffraction

As previously stated, by adjusting the magnetic lenses such that the back focal plane of the lens rather than the imaging plane is placed on the imaging apparatus a diffraction pattern can be generated. For thin crystalline samples, this produces an image that consists of a pattern of dots in the case of a single crystal, or a series of rings in the case of a polycrystalline or amorphous solid material. For the single crystal case the diffraction pattern is dependent upon the orientation of the specimen and the structure of the sample illuminated by the electron beam. This image provides the investigator with information about the space group symmetries in the crystal and the crystal's orientation to the beam path. This is typically done without utilising any information but the position at which the diffraction spots appear and the observed image symmetries.

Diffraction patterns can have a large dynamic range, and for crystalline samples, may have intensities greater than those recordable by CCD. As such, TEMs may still be equipped with film cartridges for the purpose of obtaining these images, as the film is a single use detector.

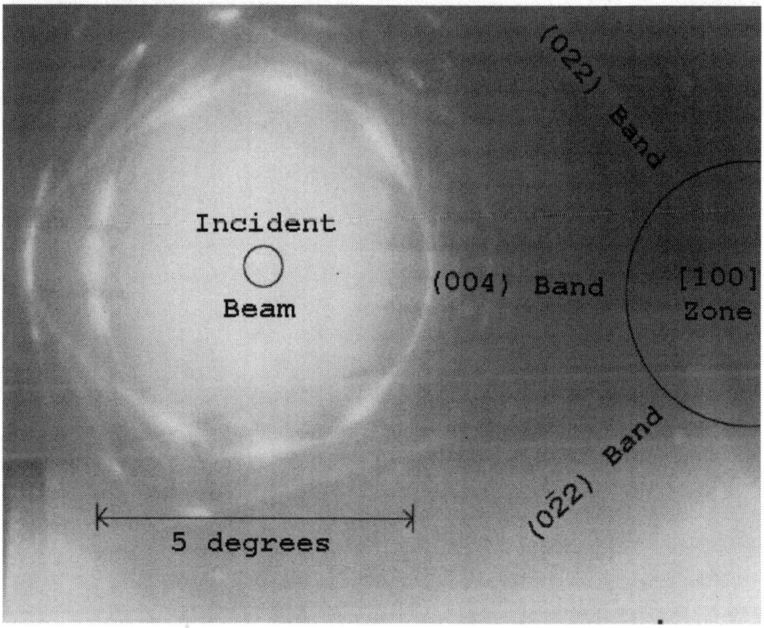

Figure: *Convergent-beam Kikuchi lines from silicon, near the [100] zone axis*

Analysis of diffraction patterns beyond point-position can be complex, as the image is sensitive to a number of factors such as specimen thickness and orientation, objective lens defocus, spherical and chromatic aberration. Although quantitative interpretation of the contrast shown in lattice images is possible, it is inherently complicated and can require extensive computer simulation and analysis, such as electron multislice analysis. More complex behaviour in the diffraction plane is also possible, with phenomena such as Kikuchi lines arising from multiple diffraction within the crystalline lattice. In convergent beam electron diffraction (CBED) where a non-parallel, i.e. converging, electron wavefront is produced by concentrating the electron beam into a fine probe at the sample surface, the interaction of the convergent beam can provide information beyond structural data such as sample thickness.

Three-Dimensional Imaging

As TEM specimen holders typically allow for the rotation of a sample by a desired angle, multiple views of the same specimen can be obtained by rotating the angle of the sample along an axis perpendicular to the beam. By taking multiple images of a single TEM sample at differing angles, typically in 1° increments, a set of images known as a "tilt series" can be collected. This methodology was proposed in the 1970s by Walter Hoppe. Under purely absorption contrast conditions, this set of images can be used to construct a three-dimensional representation of the sample.

The reconstruction is accomplished by a two-step process, first images are aligned to account for errors in the positioning of a sample; such errors can occur due to vibration or mechanical drift. Alignment methods use image registration algorithms, such as autocorrelation methods to correct these errors. Secondly, using a technique known as filtered back projection, the aligned image slices can be transformed from a set of two-dimensional images, $I_j(x, y)$, to a single three-dimensional image, $I'_j(x, y, z)$. This three-dimensional image is of particular interest when morphological information is required, further study can be undertaken using computer algorithms, such as isosurfaces and data slicing to analyse the data.

As TEM samples cannot typically be viewed at a full 180° rotation, the observed images typically suffer from a "missing wedge" of data, which when using Fourier-based back projection methods decreases the range of resolvable frequencies in the three-dimensional reconstruction. Mechanical refinements, such as multi-axis tilting (two tilt series of the same specimen made at orthogonal directions) and conical tomography (where the specimen is first tilted to a given fixed angle and then imaged at equal angular rotational increments through one complete rotation in the plane of the specimen grid) can be used to limit the impact of the missing data on the observed specimen morphology. In addition, numerical techniques exist which can improve the collected data.

All the above-mentioned methods involve recording tilt series of a given specimen field. This inevitably results in the summation of a high dose of reactive electrons through the sample and the accompanying destruction of fine detail during recording. The technique of low-dose (minimal-dose) imaging is therefore regularly applied to mitigate this effect. Low-dose imaging is performed by deflecting illumination and imaging regions simultaneously away from the optical axis to image an adjacent region to the area to be recorded (the high-dose region). This area is maintained centred during tilting and refocused before recording. During recording the deflections are removed so that the area of interest is exposed to the electron beam only for the duration required for imaging. An improvement of this technique (for objects resting on a sloping substrate film) is to have two symmetrical off-axis regions for focusing followed by setting focus to the average of the two high-dose focus values before recording the low-dose area of interest.

Non-tomographic variants on this method, referred to as single particle analysis, use images of multiple (hopefully) identical objects at different orientations to produce the image data required for three-dimensional reconstruction. If the objects do not have significant preferred orientations, this method does not suffer from the missing data wedge (or cone) which accompany tomographic methods nor does it incur excessive radiation dosage, however it assumes that the different objects imaged can be treated as if the 3D data generated from them arose from a single stable object.

Sample Preparation

Sample preparation in TEM can be a complex procedure. TEM specimens are required to be at most hundreds of nanometres thick, as unlike neutron or X-Ray radiation the electron beam interacts readily with the sample, an effect that increases roughly with atomic number squared (z^2). High quality samples will have a thickness that is comparable to the mean free path of the electrons that travel through the samples, which may be only a few tens of nanometres. Preparation of TEM specimens is specific to the material under analysis and the desired information to obtain from the specimen. As such, many generic techniques have been used for the preparation of the required thin sections.

Materials that have dimensions small enough to be electron transparent, such as powders or nanotubes, can be quickly prepared by the deposition of a dilute sample containing the specimen onto support grids or films. In the biological sciences in order to withstand the instrument vacuum and facilitate handling, biological specimens can be fixated using either a negative staining material such as uranyl acetate or by plastic embedding. Alternately samples may be held at liquid nitrogen temperatures after embedding in vitreous ice. In material science and metallurgy the specimens tend to be naturally resistant to vacuum, but still must be prepared as a thin foil, or etched so some portion of the specimen is thin enough for the beam to penetrate. Constraints on the thickness of the material may be limited by the scattering cross-section of the atoms from which the material is comprised.

Tissue Sectioning

By passing samples over a glass or diamond edge, small, thin sections can be readily obtained using a semi-automated method. This method is used to obtain thin, minimally deformed samples that allow for the observation of tissue samples. Additionally inorganic samples have been studied, such as aluminium, although this usage is limited owing to the heavy damage induced in the less soft samples. To prevent charge build-up at the sample surface, tissue samples need to be coated with a thin layer of conducting material, such as carbon, where the coating thickness is several nanometres. This may be achieved via an electric arc deposition process using a sputter coating device.

Sample Staining

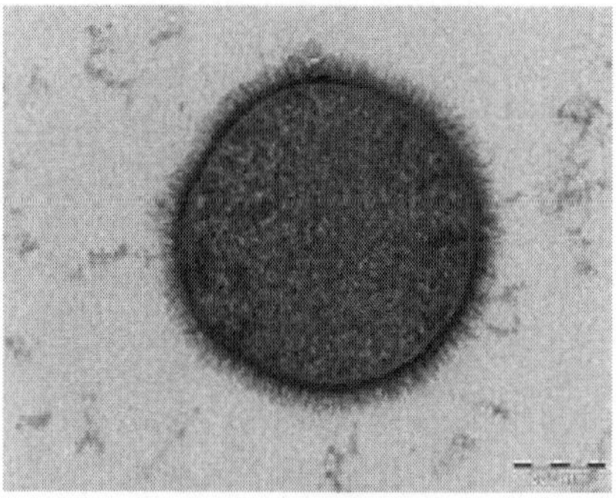

Figure: *A section of a cell of* Bacillus subtilis, *taken with a Tecnai T-12 TEM. The scale bar is 200 nm.*

Details in light microscope samples can be enhanced by stains that absorb light; similarly TEM samples of biological tissues can utilise high atomic number stains to enhance contrast. The stain absorbs electrons or scatters part of the electron beam which otherwise is projected onto the imaging system. Compounds of heavy metals such as osmium, lead, uranium or gold (in immunogold labelling) may be used prior to TEM observation to selectively deposit electron dense atoms in or on the sample in desired cellular or protein regions, requiring an understanding of how heavy metals bind to biological tissues.

Mechanical Milling

Mechanical polishing may be used to prepare samples. Polishing needs to be done to a high quality, to ensure constant sample thickness across the region of interest. A diamond, or cubic boron nitride polishing compound may be used in the final stages of polishing to remove any scratches that may cause contrast fluctuations due to varying sample thickness. Even after careful mechanical milling, additional fine methods such as ion etching may be required to perform final stage thinning.

Chemical Etching

Certain samples may be prepared by chemical etching, particularly metallic specimens. These samples are thinned using a chemical etchant, such as an acid, to prepare the sample for TEM observation. Devices to control the thinning process may allow the operator to control either the voltage or current passing through the specimen, and may include systems to detect when the sample has been thinned to a sufficient level of optical transparency.

Ion Etching

Ion etching is a sputtering process that can remove very fine quantities of material. This is used to perform a finishing polish of specimens polished by other means. Ion etching uses an inert gas passed through an electric field to generate a plasma stream that is directed to the sample surface. Acceleration energies for gases such as argon are typically a few kilovolts. The sample may be rotated to promote even polishing of the sample surface. The sputtering rate of such methods is on the order of tens of micrometres per hour, limiting the method to only extremely fine polishing.

More recently focused ion beam methods have been used to prepare samples. FIB is a relatively new technique to prepare thin samples for TEM examination from larger specimens. Because FIB can be used to micro-machine samples very precisely, it is possible to mill very thin membranes from a specific area of interest in a sample, such as a semiconductor or metal. Unlike inert gas ion sputtering, FIB makes use of significantly more energetic gallium ions and may alter the composition or structure of the material through gallium implantation.

Replication

Samples may also be replicated using cellulose acetate film, the film subsequently coated with a heavy metal, the original film melted away, and the replica imaged on the TEM. This technique is used for both materials and biological samples.

Modifications

The capabilities of the TEM can be further extended by additional stages and detectors, sometimes incorporated on the same microscope. An electron cryomicroscope *(CryoTEM) is a TEM with a specimen holder capable of maintaining the specimen at liquid nitrogen or liquid helium temperatures. This allows imaging specimens prepared in vitreous ice, the preferred preparation technique for imaging individual molecules or macromolecular assemblies.*

A TEM can be modified into a scanning transmission electron microscope (STEM) by the addition of a system that rasters the beam across the sample to form the image, combined with suitable detectors. Scanning coils are used to deflect the beam, such as by an electrostatic shift of the beam, where the beam is then collected using a current detector such as a Faraday cup, which acts as a direct electron counter. By correlating the electron count to the position of the scanning beam (known as the "probe"), the transmitted component of the beam may be measured. The non-transmitted components may be obtained either by beam tilting or by the use of annular dark field detectors.

In-situ experiments may also be conducted with experiments such as in-situ reactions or material deformation testing. Modern research TEMs may include aberration correctors, to reduce the amount of distortion in the image. Incident beam Monochromators may also be used which reduce the energy spread of the incident electron beam to less than 0.15 eV. Major TEM makers include JEOL, Hitachi High-technologies, FEI Company (from merging with Philips Electron Optics), Carl Zeiss and NION.

Low-Voltage Electron Microscope

The low-voltage electron microscope (LVEM) is a combination of SEM, TEM and STEM in one instrument, which operated at relatively low electron accelerating voltage of 5 kV. Low voltage increases image contrast which is especially important for biological specimens. This increase in contrast significantly reduces, or even eliminates the need to stain. Sectioned samples generally need to be thinner than they would be for conventional TEM (20–65 nm). Resolutions of a few nm are possible in TEM, SEM and STEM modes.

Cryo-Microscopy

This technique allows TEM's to be used to see molecular structure of proteins and large molecules. Cryoelectron microscopy involves viewing unaltered macromolecular assemblies by vitrifying them, placing them on a grid and obtaining images by detecting electrons that transmit through the specimen.

Limitations

There are a number of drawbacks to the TEM technique. Many materials require extensive sample preparation to produce a sample thin enough to be electron transparent, which makes TEM analysis a relatively time consuming process with a low throughput of samples. The structure of the sample may also be changed during the preparation process. Also the field of view is relatively small, raising the possibility that the region analyzed may not be characteristic of the whole sample. There is potential that the sample may be damaged by the electron beam, particularly in the case of biological materials.

Resolution Limits

The limit of resolution obtainable in a TEM may be described in several ways, and is typically referred to as the information limit of the microscope. One commonly used value is a cut-off value of the contrast transfer function, a function that is usually quoted in the frequency domain to define the reproduction of spatial frequencies of objects in the object plane by the microscope optics. A cut-off frequency, q_{max}, for the transfer function may be approximated with the following equation, where C_s is the spherical aberration coefficient and λ is the electron wavelength:

$$q_{\max} = \frac{1}{0.67(C_s\lambda^3)^{1/4}}.$$

For a 200 kV microscope, with partly corrected spherical aberrations ("to the third order") and a C_s value of 1 µm, a theoretical cut-off value might be $1/q_{max}$ = 42 pm. The same microscope without a corrector would have C_s = 0.5 mm and thus a 200-pm cut-off. The spherical aberrations are suppressed to the third or fifth order in the "aberration-corrected" microscopes. Their resolution is however limited by electron source geometry and brightness and chromatic aberrations in the objective lens system.

The frequency domain representation of the contrast transfer function may often have an oscillatory nature, which can be tuned by adjusting the focal value of the objective lens. This oscillatory nature implies that some spatial frequencies are faithfully imaged by the microscope, whilst others are suppressed. By combining multiple images with different spatial frequencies, the use of techniques such as focal series reconstruction can be used to improve the resolution of the TEM in a limited manner. The contrast transfer function can, to some extent, be experimentally approximated through techniques such as Fourier transforming images of amorphous material, such as amorphous carbon.

More recently, advances in aberration corrector design have been able to reduce spherical aberrations and to achieve resolution below 0.5 Ångströms (50 pm) at magnifications above 50 million times. Improved resolution allows for the imaging of lighter atoms that scatter electrons less efficiently, such as lithium atoms in lithium battery materials. The ability to determine the position of atoms within materials has made the HRTEM an indispensable tool for nanotechnology research and development in many fields, including heterogeneous catalysis and the development of semiconductor devices for electronics and photonics.

2

Rutherford Backscattering Spectrometry

Rutherford backscattering spectrometry (RBS) is an analytical technique used in materials science. Sometimes referred to as high-energy ion scattering (HEIS) spectrometry, RBS is used to determine the structure and composition of materials by measuring the backscattering of a beam of high energy ions (typically protons or alpha particles) impinging on a sample.

The Geiger–Marsden Experiment

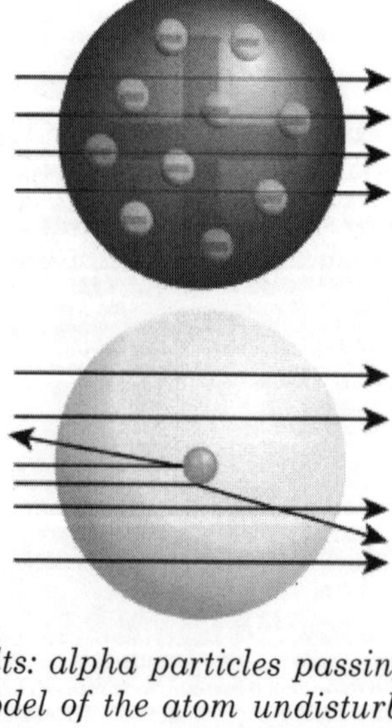

Figure: Top: *Expected results: alpha particles passing through the plum pudding model of the atom undisturbed.*

Bottom: Observed results: a small portion of the particles were deflected, indicating a small, concentrated positive charge.

Rutherford backscattering spectrometry is named after Lord Ernest Rutherford, a physicist sometimes referred to as the father of nuclear physics. Rutherford supervised a series of experiments carried out by Hans Geiger and Ernest Marsden between 1909 and 1914 studying the scattering of alpha particles through metal foils. While attempting to eliminate "stray particles" they believed to be caused by an imperfection in their alpha source, Rutherford suggested that Marsden attempt to measure backscattering from a gold foil sample. According to the then-dominant plum-pudding model of the atom, in which small negative electrons were spread through a diffuse positive region, backscattering of the high-energy positive alpha particles should have been nonexistent. At most small deflections should occur as the alpha particles passed almost unhindered through the foil. Instead, when Marsden positioned the detector on the same side of the foil as the alpha particle source, he immediately detected a noticeable backscattered signal. According to Rutherford, "It was quite the most incredible event that has ever happened to me in my life. It was almost as incredible as if you fired a 15-inch shell at a piece of tissue paper and it came back and hit you."

Rutherford interpreted the result of the Geiger–Marsden experiment as an indication of a Coulomb collision with a single massive positive particle. This led him to the conclusion that the atom's positive charge could not be diffuse but instead must be concentrated in a single massive core: the atomic nucleus. Calculations indicated that the charge necessary to accomplish this deflection was approximately 100 times the charge of the electron, close to the atomic number of gold. This led to the development of the Rutherford model of the atom in which a positive nucleus made up of Ne positive particles, or protons, was surrounded by N orbiting electrons of charge e to balance the nuclear charge. This model was eventually superseded by the Bohr atom, incorporating some early results from quantum mechanics.

If the energy of the incident particle is increased sufficiently, the Coulomb barrier is exceeded and the wavefunctions of the incident and struck particles overlap. This may result in nuclear reactions in certain cases, but frequently the interaction remains elastic, although the scattering cross-sections may fluctuate wildly as a function of energy. This case is known as "Elastic (non-Rutherford) Backscattering Spectrometry" (EBS). There has recently been enormous progress in determining EBS scattering cross-sections, by solving Schrödinger's equation for each interaction.

Basic Principles

We can describe Rutherford backscattering as an elastic (hard-sphere) collision between a high kinetic energy particle from the incident beam (the projectile*) and a stationary particle located in the sample (the* target*).* Elastic *in this context means that no energy is either lost or gained during the collision.*

Note that the "law" of the conservation of energy is not generally applicable for nuclear interactions, since in some circumstances a collision may result in a nuclear reaction, with the release of what can be very considerable quantities of energy. Nuclear reaction analysis (NRA) is very useful for detecting light elements. The energy conservation law still applies for NRA, but in the more general Mass–energy form.

Considering the kinematics of the collision (that is, the conservation of momentum and kinetic energy), the energy E_1 of the scattered projectile is reduced from the initial energy E_0:

$$E_1 = k \cdot E_0,$$

where k is known as the *kinematical factor*, and

$$k = \left(\frac{m_1 \cos\theta_1 \pm \sqrt{m_2^2 - m_1^2 (\sin\theta_1)^2}}{m_1 + m_2} \right)^2,$$

where particle 1 is the projectile, particle 2 is the target nucleus, and θ_1 is the scattering angle of the projectile in the laboratory frame of reference (that is, relative to the observer). The plus sign is taken when the mass of the projectile is less than that of the target, otherwise the minus sign is taken.

While this equation correctly determines the energy of the scattered projectile for any particular scattering angle (relative to the observer), it does not describe the probability of observing such an event. For that we need the *differential cross-section* of the backscattering event:

$$\frac{d\omega}{d\Omega} = \left(\frac{Z_1 Z_2 e^2}{4E_0} \right)^2 \frac{1}{(\sin\theta/2)^4},$$

where Z_1 and Z_2 are the atomic numbers of the incident and target nuclei. This equation is written in the centre of mass frame of reference and is therefore not a function of the mass of either the projectile or the target nucleus.

Note that the scattering angle in the laboratory frame of reference θ_1 is *not* the same as the scattering angle in the centre of mass frame of reference θ (although for RBS experiments they are usually very similar). However, heavy ion projectiles can easily recoil lighter ions which, if the geometry is right, can be ejected from the target and detected. This is the basis of the Elastic Recoil Detection (ERD, with synonyms ERDA, FRS, HFS) technique. RBS often uses a He beam which readily recoils H, so simultaneous RBS/ERD is frequently done to probe the hydrogen isotope content of samples (although H ERD with a He beam above 1 MeV is not Rutherford). For ERD the scattering angle in the lab frame of reference is quite different from that in the centre of mass frame of reference.

Note also that heavy ions cannot *back*scatter from light ones: it is kinematically prohibited. The kinematical factor must remain real, and this limits the permitted scattering angle in the laboratory frame of reference. In ERD it is often convenient to place the recoil detector at recoil angles large enough to prohibit signal from the scattered beam. The scattered ion intensity is always very large compared to the recoil intensity (the Rutherford scattering cross-section formula goes to infinity as the scattering angle goes to zero), and for ERD the scattered beam usually has to be excluded from the measurement somehow.

The singularity in the Rutherford scattering cross-section formula is unphysical of course. If the scattering cross-section is zero it implies that the projectile never comes close to the target, but in this case it also never penetrates the electron cloud surrounding the nucleus either. The pure Coulomb formula for the scattering cross-section shown above must be corrected for this screening effect, which becomes more important as the energy of the projectile decreases (or, equivalently, its mass increases).

While large-angle scattering only occurs for ions which scatter off target nuclei, inelastic small-angle scattering can also occur off the sample electrons. This results in a gradual decrease in ions which penetrate more deeply into the sample, so that backscattering off interior nuclei occurs with a lower "effective" incident energy. The amount by which the ion energy is lowered after passing through a given distance is referred to as the stopping power of the material and is dependent on the electron distribution. This energy loss varies continuously with respect to distance traversed, so that stopping power is expressed as

$$S(E) = -\frac{dE}{dx}.$$

For high energy ions stopping power is usually proportional to $\frac{Z_2}{E}$; however, precise calculation of stopping power is difficult to carry out with any accuracy.

Stopping power (properly, *stopping force*) has units of energy per unit length. It is generally given in thin film units, that is eV /(atom/cm) since it is measured experimentally on thin films whose thickness is always measured absolutely as mass per unit area, avoiding the problem of determining the density of the material which may vary as a function of thickness. Stopping power is now known for all materials at around 2%.

Instrumentation

An RBS instrument generally includes three essential components:

- An ion source, usually alpha particles (He^{2+} ions) or, less commonly, protons.

- A linear particle accelerator capable of accelerating incident ions to high energies, usually in the range 1-3 MeV.
- A detector capable of measuring the energies of backscattered ions over some range of angles.

Two common source/acceleration arrangements are used in commercial RBS systems, working in either one or two stages. One-stage systems consist of a He^+ source connected to an acceleration tube with a high positive potential applied to the ion source, and the ground at the end of the acceleration tube. This arrangement is simple and convenient, but it can be difficult to achieve energies of much more than 1 MeV due to the difficulty of applying very high voltages to the system.

Two-stage systems, or "tandem accelerators", start with a source of He^- ions and position the positive terminal at the centre of the acceleration tube. A stripper element included in the positive terminal removes electrons from ions which pass through, converting He^- ions to He^{++} ions. The ions thus start out being attracted to the terminal, pass through and become positive, and are repelled until they exit the tube at ground. This arrangement, though more complex, has the advantage of achieving higher accelerations with lower applied voltages: a typical tandem accelerator with an applied voltage of 750 kV can achieve ion energies of over 2 MeV.

Detectors to measure backscattered energy are usually silicon surface barrier detectors, a very thin layer (100 nm) of P-type silicon on an N-type substrate forming a p-n junction. Ions which reach the detector lose some of their energy to inelastic scattering from the electrons, and some of these electrons gain enough energy to overcome the band gap between the semiconductor valence and conduction bands. This means that each ion incident on the detector will produce some number of electron-hole pairs which is dependent on the energy of the ion. These pairs can be detected by applying a voltage across the detector and measuring the current, providing an effective measurement of the ion energy. The relationship between ion energy and the number of electron-hole pairs produced will be dependent on the detector materials, the type of ion and the efficiency of the current measurement; energy resolution is dependent on thermal fluctuations. It should also be noted that after one ion is incident on the detector, there will be some *dead time* before the electron-hole pairs recombine in which a second incident ion cannot be distinguished from the first.

Angular dependence of detection can be achieved by using a movable detector, or more practically by separating the surface barrier detector into many independent cells which can be measured independently, covering some range of angles around direct (180 degrees) back-scattering. Angular dependence of the incident beam is controlled by using a tiltable sample stage.

Composition and Depth Measurement

The energy loss of a backscattered ion is dependent on two processes: the energy lost in scattering events with sample nuclei, and the energy lost to small-angle scattering from the sample electrons. The first process is dependent on the scattering cross-section of the nucleus and thus on its mass and atomic number. For a given measurement angle, nuclei of two different elements will therefore scatter incident ions to different degrees and with different energies, producing separate peaks on an N(E) plot of measurement count versus energy. These peaks are characteristic of the elements contained in the material, providing a means of analyzing the composition of a sample by matching scattered energies to known scattering cross-sections. Relative concentrations can be determined by measuring the heights of the peaks.

The second energy loss process, the stopping power of the sample electrons, does not result in large discrete losses such as those produced by nuclear collisions. Instead it creates a gradual energy loss dependent on the electron density and the distance traversed in the sample. This energy loss will lower the measured energy of ions which backscatter from nuclei inside the sample in a continuous manner dependent on the depth of the nuclei. The result is that instead of the sharp backscattered peaks one would expect on an N(E) plot, with the width determined by energy and angular resolution, the peaks observed trail off gradually towards lower energy as the ions pass through the depth occupied by that element. Elements which only appear at some depth inside the sample will also have their peak positions shifted by some amount which represents the distance an ion had to traverse to reach those nuclei.

In practice, then, a compositional depth profile can be determined from an RBS N(E) measurement. The elements contained by a sample can be determined from the positions of peaks in the energy spectrum. Depth can be determined from the width and shifted position of these peaks, and relative concentration from the peak heights. This is especially useful for the analysis of a multilayer sample, for example, or for a sample with a composition which varies more continuously with depth.

This kind of measurement can only be used to determine elemental composition; the chemical structure of the sample cannot be determined from the N(E) profile. However, it is possible to learn something about this through RBS by examining the crystal structure. This kind of spatial information can be investigated by taking advantage of blocking and channeling.

Structural Measurements: Blocking and Channeling

To fully understand the interaction of an incident beam of nuclei with a crystalline structure, it is necessary to comprehend two more key concepts: blocking *and* channeling.

When a beam of ions with parallel trajectories is incident on a target atom, scattering off that atom will prevent collisions in a cone-shaped region "behind" the target relative to the beam. This occurs because the repulsive potential of the target atom bends close ion trajectories away from their original path, and is referred to as blocking. The radius of this blocked region, at a distance L from the original atom, is given by

$$R = 2\sqrt{\frac{Z_1 Z_2 e^2 L}{E_0}}$$

When an ion is scattered from deep inside a sample, it can then re-scatter off a second atom, creating a second blocked cone in the direction of the scattered trajectory. This can be detected by carefully varying the detection angle relative to the incident angle.

Channeling is observed when the incident beam is aligned with a major symmetry axis of the crystal. Incident nuclei which avoid collisions with surface atoms are excluded from collisions with all atoms deeper in the sample, due to blocking by the first layer of atoms. When the interatomic distance is large compared to the radius of the blocked cone, the incident ions can penetrate many times the interatomic distance without being backscattered. This can result in a drastic reduction of the observed backscattered signal when the incident beam is oriented along one of the symmetry directions, allowing determination of a sample's regular crystal structure. Channeling works best for very small blocking radii, i.e. for high-energy, low-atomic-number incident ions such as He^+.

The tolerance for the deviation of the ion beam angle of incidence relative to the symmetry direction depends on the blocking radius, making the allowable deviation angle proportional to

$$\sqrt{\frac{Z_1 Z_2}{E_0 d}}$$

While the intensity of an RBS peak is observed to decrease across most of its width when the beam is channeled, a narrow peak at the high-energy end of larger peak will often be observed, representing surface scattering from the first layer of atoms. The presence of this peak opens the possibility of surface sensitivity for RBS measurements.

Profiling of Displaced Atoms

In addition, channeling of ions can also be used to analyze a crystalline sample for lattice damage. If atoms within the target are displaced from their crystalline lattice site, this will result in a higher backscattering yield in relation to a perfect crystal. By comparing the spectrum from a sample being analyzed to that from a perfect crystal, and that obtained at a random (non-channeling) orientation (representative of a spectrum from an amorphous sample), it is possible to determine the extent of crystalline damage in terms of a fraction of displaced atoms. Multiplying this fraction by the density of the material when amorphous then also gives an estimate for the concentration of displaced atoms. The energy at which the increased backscattering occurs can also be used to determine the depth at which the displaced atoms are and a defect depth profile can be built up as a result.

Surface Sensitivity

While RBS is generally used to measure the bulk composition and structure of a sample, it is possible to obtain some information about the structure and composition of the sample surface. When the signal is channeled to remove the bulk signal, careful manipulation of the incident and detection angles can be used to determine the relative positions of the first few layers of atoms, taking advantage of blocking effects.

The surface structure of a sample can be changed from the ideal in a number of ways. The first layer of atoms can change its distance from subsequent layers (relaxation); it can assume a different two-dimensional structure than the bulk (reconstruction); or another material can be adsorbed onto the surface. Each of these cases can be detected by RBS. For example, surface reconstruction can be detected by aligning the beam in such a way that channeling should occur, so that only a surface peak of known intensity should be detected. A higher-than-usual intensity or a wider peak will indicate that the first layers of atoms are failing to block the layers beneath, i.e. that the surface has been reconstructed. Relaxations can be detected by a similar procedure with the sample tilted so the ion beam is incident at an angle selected so that first-layer atoms should block backscattering at a diagonal; that is, from atoms which are below and displaced from the blocking atom. A higher-than-expected backscattered yield will indicate that the first layer has been displaced relative to the second layer, or relaxed. Adsorbate materials will be detected by their different composition, changing the position of the surface peak relative to the expected position.

RBS has also been used to measure processes which affect the surface differently from the bulk by analyzing changes in the channeled surface peak. A well-known example of this is the RBS analysis of the premelting of lead surfaces by Frenken, Maree and van der Veen. In an RBS measurement of the Pb(110) surface, a well-defined surface peak which is stable at low temperatures was found to become wider and more intense as temperature increase past two-thirds of the bulk melting temperature. The peak reached the bulk height and width as temperature reached the melting temperature. This increase in the disorder of the surface, making deeper atoms visible to the incident beam, was interpreted as pre-melting of the surface, and computer simulations of the RBS process produced similar results when compared with theoretical pre-melting predictions.

RBS has also been combined with nuclear microscopy, in which a focused ion beam is scanned across a surface in a manner similar to a scanning electron microscope. The energetic analysis of backscattered signals in this kind of application provides compositional information about the surface, while the microprobe itself can be used to examine features such as periodic surface structures.

Scanning Probe Microscopy (SPM)

Scanning probe microscopy (SPM) is a branch of microscopy that forms images of surfaces using a physical probe that scans the specimen. SPM was founded with the invention of the scanning tunneling microscope in 1981.

Many scanning probe microscopes can image several interactions simultaneously. The manner of using these interactions to obtain an image is generally called a mode.

The resolution varies somewhat from technique to technique, but some probe techniques reach a rather impressive atomic resolution. They owe this largely to the ability of piezoelectric actuators to execute motions with a precision and accuracy at the atomic level or better on electronic command. One could rightly call this family of techniques "piezoelectric techniques". The other common denominator is that the data are typically obtained as a two-dimensional grid of data points, visualized in false colour as a computer image.

Established Types of Scanning Probe Microscopy

- AFM, atomic force microscopy
  - Contact AFM
  - Non-contact AFM
  - Dynamic contact AFM
  - Tapping AFM
- BEEM, ballistic electron emission microscopy
- CFM, chemical force microscopy
- C-AFM, conductive atomic force microscopy
- ECSTM electrochemical scanning tunneling microscope
- EFM, electrostatic force microscopy
- FluidFM, fluidic force microscope
- FMM, force modulation microscopy
- FOSPM, feature-oriented scanning probe microscopy

- KPFM, kelvin probe force microscopy
- MFM, magnetic force microscopy
- MRFM, magnetic resonance force microscopy
- NSOM, near-field scanning optical microscopy (or SNOM, scanning near-field optical microscopy)
- PFM, Piezoresponse Force Microscopy
- PSTM, photon scanning tunneling microscopy
- PTMS, photothermal microspectroscopy/microscopy
- SCM, scanning capacitance microscopy
- SECM, scanning electrochemical microscopy
- SGM, scanning gate microscopy
- SHPM, scanning Hall probe microscopy
- SICM, scanning ion-conductance microscopy
- SPSM spin polarized scanning tunneling microscopy
- SSRM, scanning spreading resistance microscopy
- SThM, scanning thermal microscopy
- STM, scanning tunneling microscopy
- STP, scanning tunneling potentiometry
- SVM, scanning voltage microscopy
- SXSTM, synchrotron x-ray scanning tunneling microscopy
- SSET Scanning Single-Electron Transistor Microscopy

Of these techniques AFM and STM are the most commonly used for roughness measurements.

Image Formation

To form images SPMs raster scan the tip over the surface. At discrete points in the raster scan a value is recorded (which value depends on the type of SPM and the mode of operation). These recorded values are displayed as a heat map to produce the final STM images, usually using a black and white or an orange colour scale.

Constant Interaction Mode

In constant interaction mode (often referred to as "in feedback"), a feedback loop is used to physically move the probe closer to or further from the surface (in the z axis) under study to maintain a constant interaction. This interaction depends on the type of SPM, for scanning tunneling microscopy the interaction is the tunnel current, for contact mode AFM or MFM it is the cantilever deflection, etc. The type of feedback loop used is usually a PI-loop, which is a PID-loop where the differential gain has been set to zero (as it amplifies noise). The z position of the tip (scanning plane is the xy-plane) is recorded periodically and displayed as a heat map. This is normally referred to as a topography image.

In this mode a second image, known as the 3 error signal" or "error image" is also taken, which is a heat map of the interaction which was fed back on. Under perfect operation this image would be a blank at a constant value which was set on the feedback loop. Under real operation the image shows noise and often some indication of the surface structure. The user can use this image to edit the feedback gains to minimise features in the error signal.

If the gains are set incorrectly, many imaging artifacts are possible. If gains are too low features can appear smeared. If the gains are too high the feedback can become unstable and oscillate, producing striped features in the images which are not physical.

Constant Height Mode

In constant height mode the probe is not moved in the z-axis during the raster scan. Instead the value of the interaction under study is recorded (i.e. the tunnel current for STM, or the cantilever oscillation amplitude for amplitude modulated non-contact AFM). This recorded information is displayed as a heat map, and is usually referred to as a constant height image.

Constant height imaging is much more difficult than constant interaction imaging as the probe is much more likely to crash into the sample surface. Usually before performing constant height imaging one must image in constant interaction mode to check the surface has no large contaminants in the imaging region, to measure and correct for the sample tilt, and (especially for slow scans) to measure and correct for thermal drift of the sample. Piezoelectric creep can also be a problem, so the microscope often needs time to settle after large movements before constant height imaging can be performed.

Constant height imaging can be advantageous for eliminating the possibility of feedback artifacts.

Probe Tips

The nature of an SPM probe depends entirely on the type of SPM being used. However, certain characteristics are common to all, or at least most, SPMs. Most importantly the probe must have a very sharp apex. The apex of the probe defines the resolution of the microscope, the sharper the probe the better the resolution. For atomic resolution imaging the probe must be terminated by a single atom.

For many cantilever based SPMs (e.g. AFM and MFM), the entire cantilever and integrated probe are fabricated by acid [etching], usually from silicon nitride. Conducting probes, needed for STM and SCM among others, are usually constructed from platinum/iridium wire for ambient operations, or tungsten for UHV operation. Other materials such as gold are sometimes used either for sample specific reasons or if the SPM is to be combined with other experiments such as TERS. Platinum/iridium (and other ambient) probes are normally cut using sharp wire cutters, the optimal method is to cut most of the way through the wire and then pull to snap the last of the wire, increasing the likelihood of a single atom termination. Tungsten wires are usually electrochemically etched, following this the oxide layer normally needs to be removed once the tip is in UHV conditions.

It is not uncommon for SPM probes (both purchased and "home-made") to not image with the desired resolution. This could be a tip which is too blunt or the probe may have more than one peak, resulting in a doubled or ghost image. For some probes, in situ modification of the tip apex is possible, this is usually done by either crashing the tip into the surface or by applying a large electric field. The latter is achieved by applying a bias voltage (of order 10V) between the tip and the sample, as this distance is usually 1-3 Angstroms, a very large field is generated.

Advantages of Scanning Probe Microscopy

- The resolution of the microscopes is not limited by diffraction, but only by the size of the probe-sample interaction volume (i.e., point spread function), which can be as small as a few picometres. Hence the ability to measure small local differences in object height (like that of 135 picometre steps on <100> silicon) is unparalleled. Laterally the probe-sample interaction extends only across the tip atom or atoms involved in the interaction.
- The interaction can be used to modify the sample to create small structures (nanolithography).
- Unlike electron microscope methods, specimens do not require a partial vacuum but can be observed in air at standard temperature and pressure or while submerged in a liquid reaction vessel.

Disadvantages of Scanning Probe Microscopy

- The detailed shape of the scanning tip is sometimes difficult to determine. Its effect on the resulting data is particularly noticeable if the specimen varies greatly in height over lateral distances of 10 nm or less.

- The scanning techniques are generally slower in acquiring images, due to the scanning process. As a result, efforts are being made to greatly improve the scanning rate. Like all scanning techniques, the embedding of spatial information into a time sequence opens the door to uncertainties in metrology, say of lateral spacings and angles, which arise due to time-domain effects like specimen drift, feedback loop oscillation, and mechanical vibration.
- The maximum image size is generally smaller.
- Scanning probe microscopy is often not useful for examining buried solid-solid or liquid-liquid interfaces.

Visualization and Analysis Software

In all instances and contrary to optical microscopes, rendering software is necessary to produce images. Such software is produced and embedded by instrument manufacturers but also available as an accessory from specialised work groups or companies. The main packages used are

- Freeware : Gwyddion, WSxM (developed by Nanotec).
- Commercial : FemtoScan Online (developed by Advanced Technologies Centre), SPIP (developed by ImageMet), MountainsMap SPM (developed by Digital Surf).

Scanning Tunneling Microscope (STM)

A scanning tunneling microscope (STM) is an instrument for imaging surfaces at the atomic level. Its development in 1981 earned its inventors, Gerd Binnig and Heinrich Rohrer (at IBM Zürich), the Nobel Prize in Physics in 1986. For an STM, good resolution is considered to be 0.1 nm lateral resolution and 0.01 nm depth resolution. With this resolution, individual atoms within materials are routinely imaged and manipulated. The STM can be used not only in ultra-high vacuum but also in air, water, and various other liquid or gas ambients, and at temperatures ranging from near zero kelvin to a few hundred degrees Celsius.

The STM is based on the concept of quantum tunneling. When a conducting tip is brought very near to the surface to be examined, a bias (voltage difference) applied between the two can allow electrons to tunnel through the vacuum between them. The resulting *tunneling current* is a function of tip position, applied voltage, and the local density of states (LDOS) of the sample. Information is acquired by monitoring the current as the tip's position scans across the surface, and is usually displayed in image form. STM can be a challenging technique, as it requires extremely clean and stable surfaces, sharp tips, excellent vibration control, and sophisticated electronics, but nonetheless many hobbyists have built their own.

Procedure

Figure: *A close-up of a simple scanning tunneling microscope head using a platinum–iridium tip.*

First, a voltage bias is applied and the tip is brought close to the sample by coarse sample-to-tip control, which is turned off when the tip and sample are sufficiently close. At close range, fine control of the tip in all three dimensions when near the sample is typically piezoelectric, maintaining tip-sample separation W typically in the 4-7 Å (0.4-0.7 nm) range, which is the equilibrium position between attractive (3<W<10Å) and repulsive (W<3Å) interactions. In this situation, the voltage bias will cause electrons to tunnel between the tip and sample, creating a current that can be measured. Once tunneling is established, the tip's bias and position with respect to the sample can be varied (with the details of this variation depending on the experiment) and data are obtained from the resulting changes in current.

If the tip is moved across the sample in the x-y plane, the changes in surface height and density of states cause changes in current. These changes are mapped in images. This change in current with respect to position can be measured itself, or the height, z, of the tip corresponding to a constant current can be measured. These two modes are called constant height mode and constant current mode, respectively. In constant current mode, feedback electronics adjust the height by a voltage to the piezoelectric height control mechanism. This leads to a height variation and thus the image comes from the tip topography across the sample and gives a constant charge density surface; this means contrast on the image is due to variations in charge density. In constant height mode, the voltage and height are both held constant while the current changes to keep the voltage from changing; this leads to an image made of current changes over the surface, which can be related to charge density. The benefit to using a constant height mode is that it is faster, as the piezoelectric movements require more time to register the height change in constant current mode than the current change in constant height mode.

All images produced by STM are grayscale, with colour optionally added in post-processing in order to visually emphasize important features.

In addition to scanning across the sample, information on the electronic structure at a given location in the sample can be obtained by sweeping voltage and measuring current at a specific location. This type of measurement is called scanning tunneling spectroscopy (STS) and typically results in a plot of the local density of states as a function of energy within the sample. The advantage of STM over other measurements of the density of states lies in its ability to make extremely local measurements: for example, the density of states at an impurity site can be compared to the density of states far from impurities.

Framerates of at least 1 Hz enable so called Video-STM (up to 50 Hz is possible). This can be used to scan surface diffusion.

Instrumentation

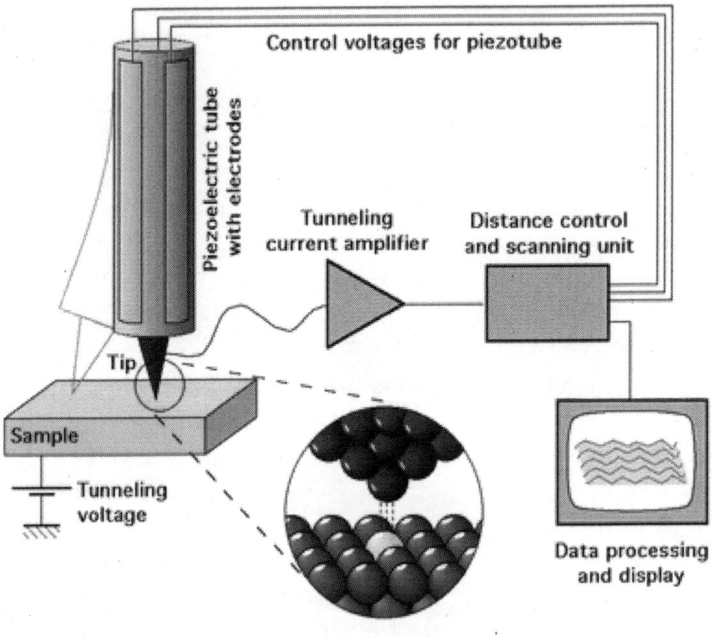

Figure: *Schematic view of an STM*

The components of an STM include scanning tip, piezoelectric controlled height and x,y scanner, coarse sample-to-tip control, vibration isolation system, and computer.

The resolution of an image is limited by the radius of curvature of the scanning tip of the STM. Additionally, image artifacts can occur if the tip has two tips at the end rather than a single atom; this leads to "double-tip imaging," a situation in which both tips contribute to the tunneling. Therefore it has been essential to develop processes for consistently obtaining sharp, usable tips. Recently, carbon nanotubes have been used in this instance.

The tip is often made of tungsten or platinum-iridium, though gold is also used. Tungsten tips are usually made by electrochemical etching, and platinum-iridium tips by mechanical shearing.

Due to the extreme sensitivity of tunnel current to height, proper vibration isolation or an extremely rigid STM body is imperative for obtaining usable results. In the first STM by Binnig and Rohrer, magnetic levitation was used to keep the STM free from vibrations; now mechanical spring or gas spring systems are often used. Additionally, mechanisms for reducing eddy currents are sometimes implemented.

Maintaining the tip position with respect to the sample, scanning the sample and acquiring the data is computer controlled. The computer may also be used for enhancing the image with the help of image processing as well as performing quantitative measurements.

Probe Tips

STM tips are usually made from W (tungsten) metal or Pt/Ir alloy where at the very end of the tip (called apex), there is one atom of the material.

Other STM Related Studies

Many other microscopy techniques have been developed based upon STM. These include photon scanning microscopy (PSTM), which uses an optical tip to tunnel photons; scanning tunneling potentiometry (STP), which measures electric potential across a surface; spin polarized scanning tunneling microscopy (SPSTM), which uses a ferromagnetic tip to tunnel spin-polarized electrons into a magnetic sample, and atomic force microscopy (AFM), in which the force caused by interaction between the tip and sample is measured.

Other STM methods involve manipulating the tip in order to change the topography of the sample. This is attractive for several reasons. Firstly the STM has an atomically precise positioning system which allows very accurate atomic scale manipulation. Furthermore, after the surface is modified by the tip, it is a simple matter to then image with the same tip, without changing the instrument. IBM researchers developed a way to manipulate xenon atoms adsorbed on a nickel surface. This technique has been used to create electron "corrals" with a small number of adsorbed atoms, which allows the STM to be used to observe electron Friedel oscillations on the surface of the material. Aside from modifying the actual sample surface, one can also use the STM to tunnel electrons into a layer of electron beam photoresist on a sample, in order to do lithography. This has the advantage of offering more control of the exposure than traditional electron beam lithography. Another practical application of STM is atomic deposition of metals (Au, Ag, W, etc.) with any desired (pre-programmed) pattern, which can be used as contacts to nanodevices or as nanodevices themselves.

Recently groups have found they can use the STM tip to rotate individual bonds within single molecules. The electrical resistance of the molecule depends on the orientation of the bond, so the molecule effectively becomes a molecular switch.

Principle of Operation

Tunneling is a functioning concept that arises from quantum mechanics. Classically, an object hitting an impenetrable barrier will not pass through. In contrast, objects with a very small mass, such as the electron, have wavelike characteristics which permit such an event, referred to as tunneling.

Electrons behave as beams of energy, and in the presence of a potential $U(z)$, assuming 1-dimensional case, the energy levels $\psi_n(z)$ of the electrons are given by solutions to Schrödinger's equation,

$$-\frac{\hbar^2}{2m}\frac{\partial^2\psi_n(z)}{\partial z^2}+U(z)\psi_n(z)=E\psi_n(z)$$

where ' is the reduced Planck's constant, z is the position, and m is the mass of an electron. If an electron of energy E is incident upon an energy barrier of height $U(z)$, the electron wave function is a travelling wave solution,

$$\psi_n(z)=\psi_n(0)e^{\pm ikz}$$

where

$$k=\frac{\sqrt{2m(E-U(z))}}{\hbar}$$

if $E > U(z)$, which is true for a wave function inside the tip or inside the sample. Inside a barrier, $E < U(z)$ so the wave functions which satisfy this are decaying waves,

$$\psi_n(z)=\psi_n(0)e^{\pm\kappa z}$$

where

$$\kappa=\frac{\sqrt{2m(U-E)}}{\hbar}$$

quantifies the decay of the wave inside the barrier, with the barrier in the $+z$ direction for $-\kappa$.

Knowing the wave function allows one to calculate the probability density for that electron to be found at some location. In the case of tunneling, the tip and sample wave functions overlap such that when under a bias, there is some finite probability to find the electron in the barrier region and even on the other side of the barrier. Let us assume the bias is V and the barrier width is W. This probability, P, that an electron at z=0 (left edge of barrier) can be found at z=W (right edge of barrier) is proportional to the wave function squared,

$$P \propto |\psi_n(0)|^2 \, e^{-2\kappa W}.$$

If the bias is small, we can let U " E H" φM in the expression for κ, where φM, the work function, gives the minimum energy needed to bring an electron from an occupied level, the highest of which is at the Fermi level (for metals at T=0 kelvins), to vacuum level. When a small bias V is applied to the system, only electronic states very near the Fermi level, within eV (a product of electron charge and voltage, not to be confused here with electronvolt unit), are excited. These excited electrons can tunnel across the barrier. In other words, tunneling occurs mainly with electrons of energies near the Fermi level.

However, tunneling does require that there is an empty level of the same energy as the electron for the electron to tunnel into on the other side of the barrier. It is because of this restriction that the tunneling current can be related to the density of available or filled states in the sample. The current due to an applied voltage V (assume tunneling occurs sample to tip) depends on two factors: 1) the number of electrons between E_f and eV in the sample, and 2) the number among them which have corresponding free states to tunnel into on the other side of the barrier at the tip. The higher density of available states the greater the tunneling current. When V is positive, electrons in the tip tunnel into empty states in the sample; for a negative bias, electrons tunnel out of occupied states in the sample into the tip.

Mathematically, this tunneling current is given by

$$I \propto \sum_{E_f - eV}^{E_f} |\psi_n(0)|^2 \, e^{-2\kappa W}.$$

One can sum the probability over energies between $E_f - eV$ and E_f to get the number of states available in this energy range per unit volume, thereby finding the local density of states (LDOS) near the Fermi level. The LDOS near some energy E in an interval ε is given by

$$\rho_s(z,E) = \frac{1}{\epsilon} \sum_{E-\epsilon}^{E} |\psi_n(z)|^2,$$

and the tunnel current at a small bias V is proportional to the LDOS near the Fermi level, which gives important information about the sample. It is desirable to use LDOS to express the current because this value does not change as the volume changes, while probability density does. Thus the tunneling current is given by

$$I \propto V\rho_s(0,E_f)e^{-2\kappa W}$$

where $\rho_s(0,E_f)$ is the LDOS near the Fermi level of the sample at the sample surface. This current can also be expressed in terms of the LDOS near the Fermi level of the sample at the tip surface,

$$I \propto V\rho_s(W,E_f)$$

The exponential term in the above equations means that small variations in W greatly influence the tunnel current. If the separation is decreased by 1 ú, the current increases by an order of magnitude, and vice versa.

This approach fails to account for the *rate* at which electrons can pass the barrier. This rate should affect the tunnel current, so it can be treated using the Fermi's golden rule with the appropriate tunneling matrix element. John Bardeen solved this problem in his study of the metal-insulator-metal junction. He found that if he solved Schrödinger's equation for each side of the junction separately to obtain the wave functions ψ and χ for each electrode, he could obtain the tunnel matrix, M, from the overlap of these two wave functions. This can be applied to STM by making the electrodes the tip and sample, assigning ψ and χ as sample and tip wave functions, respectively, and evaluating M at some surface S between the metal electrodes, where z=0 at the sample surface and z=W at the tip surface.

Now, Fermi's Golden Rule gives the rate for electron transfer across the barrier, and is written

$$w = \frac{2\pi}{\hbar}|M|^2\,\delta(E_\psi - E_\chi),$$

where $\delta(E_\psi - E_\chi)$ restricts tunneling to occur only between electron levels with the same energy. The tunnel matrix element, given by

$$M = \frac{\hbar^2}{2m}\int_{z=z_0}\left(\chi^*\frac{\partial\psi}{\partial z} - \psi\frac{\partial\chi^*}{\partial z}\right)dS,$$

is a description of the lower energy associated with the interaction of wave functions at the overlap, also called the resonance energy.

Summing over all the states gives the tunneling current as

where f is the Fermi function, ρ_s and ρ_T are the density of states in the sample and tip, respectively. The Fermi distribution function describes the filling of electron levels at a given temperature T.

Early Invention

An earlier, similar invention, the Topografiner of R. Young, J. Ward, and F. Scire from the NIST, relied on field emission. However, Young is credited by the Nobel Committee as the person who realised that it should be possible to achieve better resolution by using the tunnel effect.

Atomic Force Microscope (AFM)

Atomic force microscopy (AFM) or scanning force microscopy (SFM) is a very high-resolution type of scanning probe microscopy, with demonstrated resolution on the order of fractions of a nanometre, more than 1000 times better than the optical diffraction limit. The precursor to the AFM, the scanning tunneling microscope, was developed by Gerd Binnig and Heinrich Rohrer in the early 1980s at IBM Research - Zurich, a development that earned them the Nobel Prize for Physics in 1986. Binnig, Quate and Gerber invented the first atomic force microscope (also abbreviated as AFM) in 1986. The first commercially available atomic force microscope was introduced in 1989. The AFM is one of the foremost tools for imaging, measuring, and manipulating matter at the nanoscale. The information is gathered by "feeling" the surface with a mechanical probe. Piezoelectric elements that facilitate tiny but accurate and precise movements on (electronic) command enable the very precise scanning. In some variations, electric potentials can also be scanned using conducting cantilevers. In more advanced versions, currents can be passed through the tip to probe the electrical conductivity or transport of the underlying surface, but this is much more challenging with few research groups reporting consistent data.

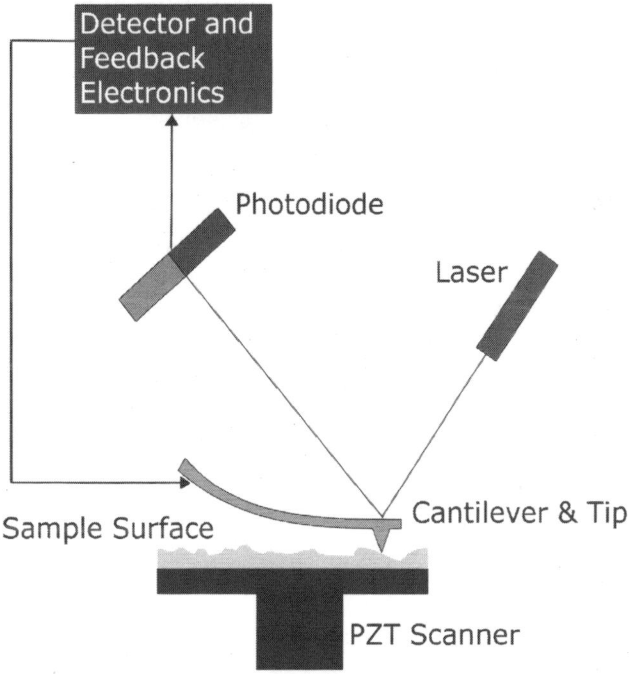

Figure: *Block diagram of atomic force microscope using beam deflection detection. As the cantilever is displaced via its interaction with the surface, so too will the reflection of the laser beam be displaced on the surface of the photodiode.*

Basic Principles

The AFM consists of a cantilever with a sharp tip (probe) at its end that is used to scan the specimen surface. The cantilever is typically silicon or silicon nitride with a tip radius of curvature on the order of nanometres. When the tip is brought into proximity of a sample surface, forces between the tip and the sample lead to a deflection of the cantilever according to Hooke's law. Depending on the situation, forces that are measured in AFM include mechanical contact force, van der Waals forces, capillary forces, chemical bonding, electrostatic forces, magnetic forces, Casimir forces, solvation forces, etc. Along with force, additional quantities may simultaneously be measured through the use of specialised types of probes. Typically, the deflection is measured using a laser spot reflected from the top surface of the cantilever into an array of photodiodes. Other methods that are used include optical interferometry, capacitive sensing or piezoresistive AFM cantilevers. These cantilevers are fabricated with piezoresistive elements that act as a strain gauge. Using a Wheatstone bridge, strain in the AFM cantilever due to deflection can be measured, but this method is not as sensitive as laser deflection or interferometry.

If the tip was scanned at a constant height, a risk would exist that the tip collides with the surface, causing damage. Hence, in most cases a feedback mechanism is employed to adjust the tip-to-sample distance to maintain a constant force between the tip and the sample. Traditionally the tip or sample is mounted on a 'tripod' of three piezo crystals, with each responsible for scanning in the *x*,*y* and *z* directions. In 1986, the same year as the AFM was invented, a new piezoelectric scanner, the tube scanner, was developed for use in STM. Later tube scanners were incorporated into AFMs. The tube scanner can move the sample in the *x*, *y*, and *z* directions using a single tube piezo with a single interior contact and four external contacts. An advantage of the tube scanner is better vibrational isolation, resulting from the higher resonant frequency of the single-crystal construction in combination with a low resonant frequency isolation stage. A disadvantage is that the *x*-*y* motion can cause unwanted *z* motion resulting in distortion.

The AFM can be operated in a number of modes, depending on the application. In general, possible imaging modes are divided into static (also called *contact*) modes and a variety of dynamic (non-contact or "tapping") modes where the cantilever is vibrated.

Probe

An AFM probe is a consumable measuring device with a sharp tip on the free-swinging end of a cantilever that is protruding from a holder plate. The dimensions of the cantilever are in the scale of micrometres. The radius of the tip is in the scale of a few nanometres. The holder plate, also called holder chip, - often 1.6 mm by 3.4 mm in size - allows the operator to hold the AFM probe with tweezers and fit it into the corresponding holder clips on the scanning head of the Atomic force microscope.

This device is most commonly called an "AFM probe", but other names include "AFM tip" and "cantilever" (employing the name of a single part as the name of the whole device). An AFM probe is a particular type of SPM (Scanning probe microscopy) probe.

AFM probes are manufactured with MEMS technology. Most AFM probes used are made from silicon (Si), but borosilicate glass and silicon nitride are also in use. As the tip is brought very close to the surface of the object under investigation, the cantilever is deflected by the interaction between the tip and the surface, which is the what the AFM is designed to measure. A spatial map of the interaction can be made by measuring the deflection at many points of a 2D surface.

Several types of interaction can be detected. Depending on the interaction under investigation, the surface of the tip of the AFM probe needs to be modified with a coating. Among the coatings used are gold - for covalent bonding of biological molecules and the detection of their interaction with a surface, diamond for increased wear resistance and magnetic coatings for detecting the magnetic properties of the investigated surface.

The surface of the cantilevers can also be modified. These coatings are mostly applied in order to increase the reflectance of the cantilever and to improve the deflection signal.

Imaging Modes

AFM operation is usually described as one of three modes, according to the nature of the tip motion:

- contact mode, also called static mode (as opposed to the other two modes, which are called dynamic modes)
- tapping mode, also called intermittent contact, ACmode, or vibrating mode, or, after the detection mechanism, Amplitude Modulation AFM
- non-contact mode, or, again after the detection mechanism, Frequency Modulation AFM

Contact Mode

In contact mode, the tip is "dragged" across the surface of the sample and the contours of the surface are measured either using the deflection of the cantilever directly or, more commonly, using the feedback signal required to keep the cantilever at a constant position. Because the measurement of a static signal is prone to noise and drift, low stiffness cantilevers are used to boost the deflection signal. Close to the surface of the sample, attractive forces can be quite strong, causing the tip to "snap-in" to the surface. Thus, contact mode AFM is almost always done at a depth where the overall force is repulsive, that is, in firm "contact" with the solid surface below any adsorbed layers.

Tapping Mode

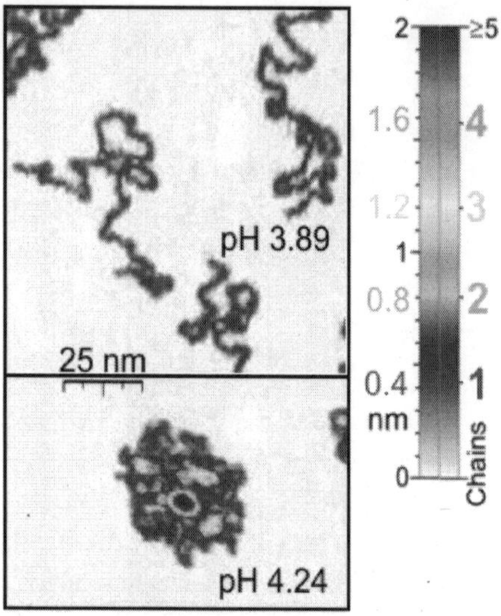

Figure: *Single polymer chains (0.4 nm thick) recorded in a tapping mode under aqueous media with different pH.*

In ambient conditions, most samples develop a liquid meniscus layer. Because of this, keeping the probe tip close enough to the sample for short-range forces to become detectable while preventing the tip from sticking to the surface presents a major problem for non-contact dynamic mode in ambient conditions. Dynamic contact mode (also called intermittent contact, AC mode or tapping mode) was developed to bypass this problem.

In *tapping mode*, the cantilever is driven to oscillate up and down at near its resonance frequency by a small piezoelectric element mounted in the AFM tip holder similar to non-contact mode. However, the amplitude of this oscillation is greater than 10 nm, typically 100 to 200 nm. The interaction of forces acting on the cantilever when the tip comes close to the surface, Van der Waals forces, dipole-dipole interactions, electrostatic forces, etc. cause the amplitude of this oscillation to decrease as the tip gets closer to the sample. An electronic servo uses the piezoelectric actuator to control the height of the cantilever above the sample. The servo adjusts the height to maintain a set cantilever oscillation amplitude as the cantilever is scanned over the sample. A *tapping AFM* image is therefore produced by imaging the force of the intermittent contacts of the tip with the sample surface.

This method of "tapping" lessens the damage done to the surface and the tip compared to the amount done in contact mode. Tapping mode is gentle enough even for the visualization of supported lipid bilayers or adsorbed single polymer molecules (for instance, 0.4 nm thick chains of synthetic polyelectrolytes) under liquid medium. With proper scanning parameters, the conformation of single molecules can remain unchanged for hours.

Non-Contact Mode

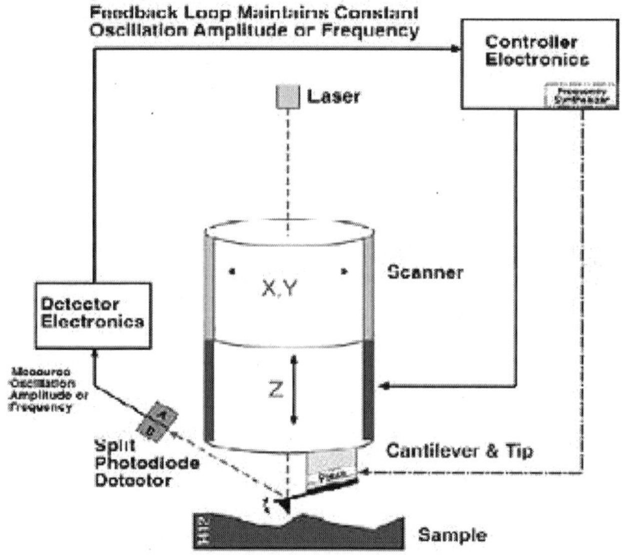

Figure: *AFM – non-contact mode*

In this mode, the tip of the cantilever does not contact the sample surface. The cantilever is instead oscillated at either its resonant frequency (frequency modulation) or just above (amplitude modulation) where the amplitude of oscillation is typically a few nanometres (<10 nm) down to a few picometres. The van der Waals forces, which are strongest from 1 nm to 10 nm above the surface, or any other long-range force that extends above the surface acts to decrease the resonance frequency of the cantilever. This decrease in resonant frequency combined with the feedback loop system maintains a constant oscillation amplitude or frequency by adjusting the average tip-to-sample distance. Measuring the tip-to-sample distance at each (x,y) data point allows the scanning software to construct a topographic image of the sample surface.

Non-contact mode AFM does not suffer from tip or sample degradation effects that are sometimes observed after taking numerous scans with contact AFM. This makes non-contact AFM preferable to contact AFM for measuring soft samples, e.g., biology sample and organic thin film. In the case of rigid samples, contact and non-contact images may look the same. However, if a few monolayers of adsorbed fluid are lying on the surface of a rigid sample, the images may look quite different. An AFM operating in contact mode will penetrate the liquid layer to image the underlying surface, whereas in non-contact mode an AFM will oscillate above the adsorbed fluid layer to image both the liquid and surface.

Schemes for dynamic mode operation include frequency modulation and the more common amplitude modulation. In frequency modulation, changes in the oscillation frequency provide information about tip-sample interactions. Frequency can be measured with very high sensitivity and thus the frequency modulation mode allows for the use of very stiff cantilevers. Stiff cantilevers provide stability very close to the surface and, as a result, this technique was the first AFM technique to provide true atomic resolution in ultra-high vacuum conditions.

In amplitude modulation, changes in the oscillation amplitude or phase provide the feedback signal for imaging. In amplitude modulation, changes in the phase of oscillation can be used to discriminate between different types of materials on the surface. Amplitude modulation can be operated either in the non-contact or in the intermittent contact regime. In dynamic contact mode, the cantilever is oscillated such that the separation distance between the cantilever tip and the sample surface is modulated.

Amplitude modulation has also been used in the non-contact regime to image with atomic resolution by using very stiff cantilevers and small amplitudes in an ultra-high vacuum environment.

AFM Cantilever Deflection Measurement

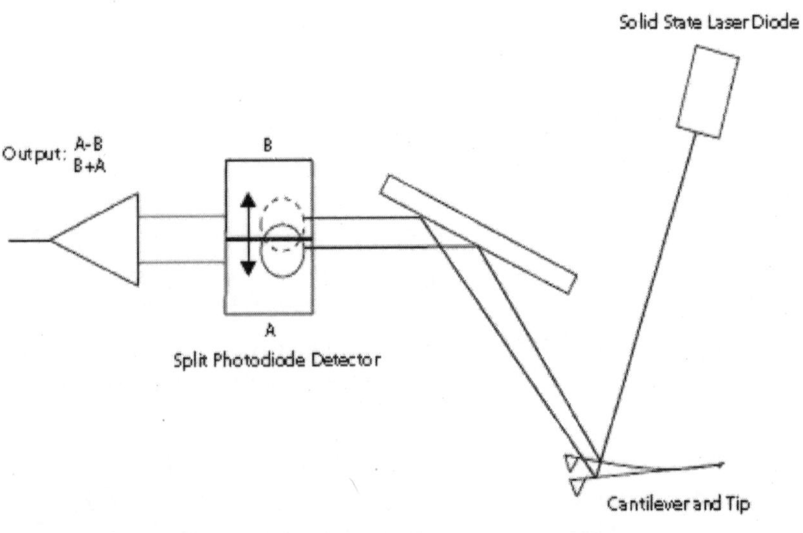

Figure: *AFM beam deflection detection*

Beam Deflection Measurement: The most common method for cantilever deflection measurements is the beam deflection method. In this method, laser light from a solid-state diode is reflected off the back of the cantilever and collected by a position-sensitive detector (PSD) consisting of two closely spaced photodiodes whose output signal is collected by a differential amplifier. Angular displacement of the cantilever results in one photodiode's collecting more light than the other photodiode, producing an output signal (the difference between the photodiode signals normalized by their sum), which is proportional to the deflection of the cantilever. It detects cantilever deflections <10 nm (thermal noise limited). A long beam path (several centimetres) amplifies changes in beam angle.

Other Deflection Measurement Methods

Many other methods for beam deflection measurements exist.

- Piezoelectric detection — Cantilevers made from quartz (such as the qPlus configuration), or other piezoelectric materials can directly detect deflection as an electrical signal. Cantilever oscillations down to 10pm have been detected with this method.
- Laser Doppler vibrometry — A laser Doppler vibrometre can be used to produce very accurate deflection measurements for an oscillating cantilever (thus is only used in non-contact mode). This method is expensive and is only used by relatively few groups.
- STM — The first atomic microscope used an STM complete with its own feedback mechanism to measure deflection. This method is very difficult to implement, and is slow to react to deflection changes compared to modern methods.
- Optical Interferometry — Optical interferometry can be used to measure cantilever deflection. Due to the nanometre scale deflections measured in AFM, the interferometre is running in the sub-fringe regime, thus, any drift in laser power or wavelength has strong effects on the measurement. For these reasons optical interferometre measurements must be done with great care (for example using index matching fluids between optical fibre junctions), with very stable lasers. For these reasons optical interferometry is rarely used.
- Capacitive detection — Metal coated cantilevers can form a capacitor with another contact located behind the cantilever. Deflection changes the distance between the contacts and can be measured as a change in capacitance.
- Piezoresistive detection — Similar to piezoelectric detection, but uses piezoresistive cantilevers to measure the detection. This is not commonly used as the piezoresistive detection dissipates energy from the system affecting Q of the resonance.

Force Spectroscopy

Another major application of AFM (besides imaging) is force spectroscopy, the direct measurement of tip-sample interaction forces as a function of the gap between the tip and sample (the result of this measurement is called a force-distance curve). For this method, the AFM tip is extended towards and retracted from the surface as the deflection of the cantilever is monitored as a function of piezoelectric displacement. These measurements have been used to measure nanoscale contacts, atomic bonding, Van der Waals forces, and Casimir forces, dissolution forces in liquids and single molecule stretching and rupture forces. Furthermore, AFM was used to measure, in an aqueous environment, the dispersion force due to polymer adsorbed on the substrate. Forces of the order of a few piconewtons can now be routinely measured with a vertical distance resolution of better than 0.1 nanometres. Force spectroscopy can be performed with either static or dynamic modes. In dynamic modes, information about the cantilever vibration is monitored in addition to the static deflection.

Problems with the technique include no direct measurement of the tip-sample separation and the common need for low-stiffness cantilevers, which tend to 'snap' to the surface. These problems are not insurmountable. An AFM that directly measures the tip-sample separation has been developed. The snap-in can be reduced by measuring in liquids or by using stiffer cantilevers, but in the latter case a more sensitive deflection sensor is needed. By applying a small dither to the tip, the stiffness (force gradient) of the bond can be measured as well.

Biological Applications

Force spectroscopy is used in biophysics to measure the mechanical properties of living material (such as tissue or cells).

Identification of Individual Surface Atoms

The AFM can be used to image and manipulate atoms and structures on a variety of surfaces. The atom at the apex of the tip "senses" individual atoms on the underlying surface when it forms incipient chemical bonds with each atom. Because these chemical interactions subtly alter the tip's vibration frequency, they can be detected and mapped. This principle was used to distinguish between atoms of silicon, tin and lead on an alloy surface, by comparing these 'atomic fingerprints' to values obtained from large-scale density functional theory (DFT) simulations.

The trick is to first measure these forces precisely for each type of atom expected in the sample, and then to compare with forces given by DFT simulations. The team found that the tip interacted most strongly with silicon atoms, and interacted 23% and 41% less strongly with tin and lead atoms, respectively. Thus, each different type of atom can be identified in the matrix as the tip is moved across the surface.

Advantages and Disadvantages

Just like any other tool, an AFM's usefulness has limitations. When determining whether or not analyzing a sample with an AFM is appropriate, there are various advantages and disadvantages that must be considered.

Advantages

AFM has several advantages over the scanning electron microscope (SEM). Unlike the electron microscope, which provides a two-dimensional projection or a two-dimensional image of a sample, the AFM provides a three-dimensional surface profile. In addition, samples viewed by AFM do not require any special treatments (such as metal/carbon coatings) that would irreversibly change or damage the sample, and does not typically suffer from charging artifacts in the final image. While an electron microscope needs an expensive vacuum environment for proper operation, most AFM modes can work perfectly well in ambient air or even a liquid environment. This makes it possible to study biological macromolecules and even living organisms. In principle, AFM can provide higher resolution than SEM. It has been shown to give true atomic resolution in ultra-high vacuum (UHV) and, more recently, in liquid environments. High resolution AFM is comparable in resolution to scanning tunneling microscopy and transmission electron microscopy. AFM can also be combined with a variety of optical microscopy techniques such as fluorescent microscopy, further expanding its applicability. Combined AFM-optical instruments have been applied primarily in the biological sciences but have also found a niche in some materials applications, especially those involving photovoltaics research.

Disadvantages

A disadvantage of AFM compared with the scanning electron microscope (SEM) is the single scan image size. In one pass, the SEM can image an area on the order of square millimetres with a depth of field on the order of millimetres, whereas the AFM can only image a maximum height on the order of 10-20 micrometres and a maximum scanning area of about 150×150 micrometres. One method of improving the scanned area size for AFM is by using parallel probes in a fashion similar to that of millipede data storage.

The scanning speed of an AFM is also a limitation. Traditionally, an AFM cannot scan images as fast as a SEM, requiring several minutes for a typical scan, while a SEM is capable of scanning at near real-time, although at relatively low quality. The relatively slow rate of scanning during AFM imaging often leads to thermal drift in the image making the AFM less suited for measuring accurate distances between topographical features on the image. However, several fast-acting designs were suggested to increase microscope scanning productivity including what is being termed videoAFM (reasonable quality images are being obtained with videoAFM at video rate: faster than the average SEM). To eliminate image distortions induced by thermal drift, several methods have been introduced.

AFM images can also be affected by nonlinearity, hysteresis, and creep of the piezoelectric material and cross-talk between the *x*, *y*, *z* axes that may require software enhancement and filtering. Such filtering could "flatten" out real topographical features. However, newer AFMs utilise real-time correction software (for example, feature-oriented scanning) or closed-loop scanners, which practically eliminate these problems. Some AFMs also use separated orthogonal scanners (as opposed to a single tube), which also serve to eliminate part of the cross-talk problems.

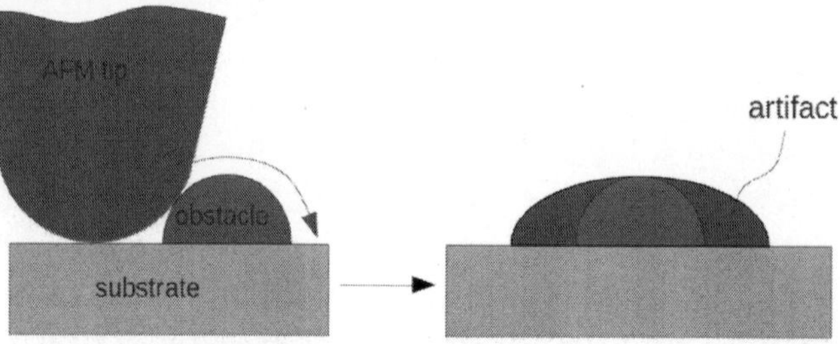

Figure: *Showing an AFM artifact arising from a tip with a high radius of curvature with respect to the feature that is to be visualized.*

As with any other imaging technique, there is the possibility of image artifacts, which could be induced by an unsuitable tip, a poor operating environment, or even by the sample itself, as depicted on the right. These image artifacts are unavoidable; however, their occurrence and effect on results can be reduced through various methods. Artifacts resulting from a too-coarse tip can be caused for example by inappropriate handling or de facto collisions with the sample by either scanning too fast or having an unreasonably rough surface, causing actual wearing of the tip.

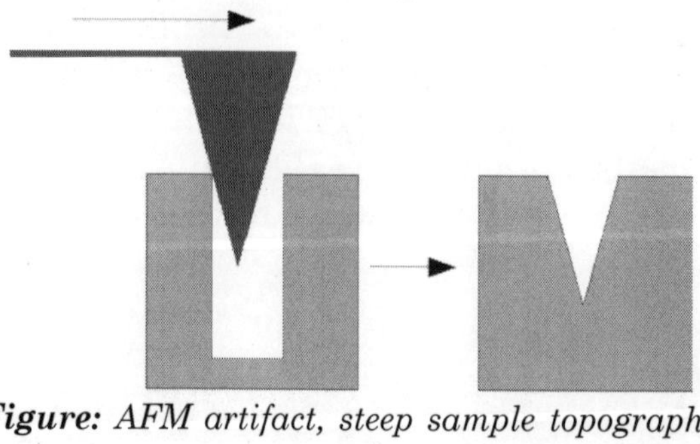

Figure: *AFM artifact, steep sample topography*

Due to the nature of AFM probes, they cannot normally measure steep walls or overhangs. Specially made cantilevers and AFMs can be used to modulate the probe sideways as well as up and down (as with dynamic contact and non-contact modes) to measure sidewalls, at the cost of more expensive cantilevers, lower lateral resolution and additional artifacts.

Piezoelectric Scanners

AFM scanners are made from piezoelectric material, which expands and contracts proportionally to an applied voltage. Whether they elongate or contract depends upon the polarity of the voltage applied. The scanner is constructed by combining independently operated piezo electrodes for X, Y, and Z into a single tube, forming a scanner that can manipulate samples and probes with extreme precision in 3 dimensions. Independent stacks of piezos can be used instead of a tube, resulting in decoupled X, Y, and Z movement.

Scanners are characterized by their sensitivity, which is the ratio of piezo movement to piezo voltage, i.e., by how much the piezo material extends or contracts per applied volt. Because of differences in material or size, the sensitivity varies from scanner to scanner. Sensitivity varies non-linearly with respect to scan size. Piezo scanners exhibit more sensitivity at the end than at the beginning of a scan. This causes the forward and reverse scans to behave differently and display hysteresis between the two scan directions. This can be corrected by applying a non-linear voltage to the piezo electrodes to cause linear scanner movement and calibrating the scanner accordingly. One disadvantage of this approach is that it requires re-calibration because the precise non-linear voltage needed to correct non-linear movement will change as the piezo ages. This problem can be circumvented by adding a linear sensor to the sample stage or piezo stage to detect the true movement of the piezo. Deviations from ideal movement can be detected by the sensor and corrections applied to the piezo drive signal to correct for non-linear piezo movement. This design is known as a 'closed loop' AFM. Non-sensored piezo AFMs are referred to as 'open loop' AFMs.

The sensitivity of piezoelectric materials decreases exponentially with time. This causes most of the change in sensitivity to occur in the initial stages of the scanner's life. Piezoelectric scanners are run for approximately 48 hours before they are shipped from the factory so that they are past the point where they may have large changes in sensitivity. As the scanner ages, the sensitivity will change less with time and the scanner would seldom require recalibration, though various manufacturer manuals recommend monthly to semi-monthly calibration of open loop AFMs.

3

Mass Spectrometry

Mass spectrometry (MS) is an analytical technique that produces spectra (singular *spectrum*) of the masses of the atoms or molecules comprising a sample of material. The spectra are used to determine the elemental or isotopic signature of a sample, the masses of particles and of molecules, and to elucidate the chemical structures of molecules, such as peptides and other chemical compounds. Mass spectrometry works by ionizing chemical compounds to generate charged molecules or molecule fragments and measuring their mass-to-charge ratios.

In a typical MS procedure, a sample, which may be solid, liquid, or gas, is ionized, for example by bombarding it with electrons. This may cause some of the sample's molecules to break into charged fragments. These ions are then separated according to their mass-to-charge ratio, typically by accelerating them and subjecting them to an electric or magnetic field: ions of the same mass-to-charge ratio will undergo the same amount of deflection. The ions are detected by a mechanism capable of detecting charged particles, such as an electron multiplier. Results are displayed as spectra of the relative abundance of detected ions as a function of the mass-to-charge ratio. The atoms or molecules in the sample can be identified by correlating known masses to the identified masses or through a characteristic fragmentation pattern.

History

In 1886, Eugen Goldstein observed rays in gas discharges under low pressure that travelled away from the anode and through channels in a perforated cathode, opposite to the direction of negatively charged cathode rays (which travel from cathode to anode). Goldstein called these positively charged anode rays "Kanalstrahlen"; the standard translation of this term into English is "canal rays". Wilhelm Wien found that strong electric or magnetic fields deflected the canal rays and, in 1899, constructed a device with parallel electric and magnetic fields that separated the positive rays according to their charge-to-mass ratio (Q/m). Wien found that the charge-to-mass ratio depended on the nature of the gas in the discharge tube. English scientist J.J. Thomson later improved on the work of Wien by reducing the pressure to create the mass spectrograph.

Some of the modern techniques of mass spectrometry were devised by Arthur Jeffrey Dempster and F.W. Aston in 1918 and 1919 respectively. In 1989, half of the Nobel Prize in Physics was awarded to Hans Dehmelt and Wolfgang Paul for the development of the ion trap technique in the 1950s and 1960s. In 2002, the Nobel Prize in Chemistry was awarded to John Bennett Fenn for the development of electrospray ionization (ESI) and Koichi Tanaka for the development of soft laser desorption (SLD) and their application to the ionization of biological macromolecules, especially proteins.

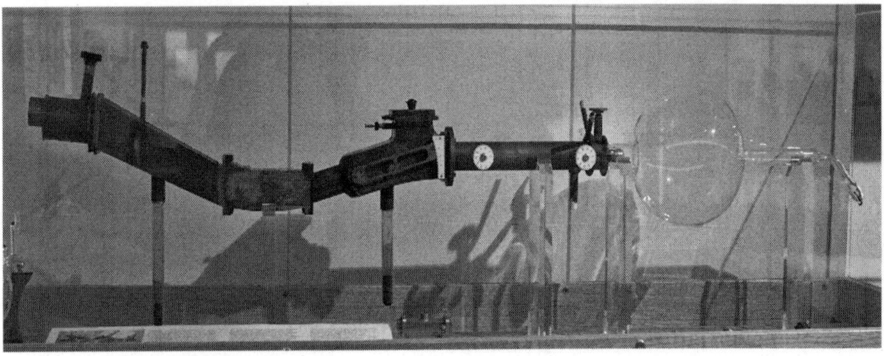

Figure: *Replica of an early mass spectrometre*

Parts of a Mass Spectrometre

A mass spectrometre consists of three components: an ion source, a mass analyzer, and a detector. The *ionizer* converts a portion of the sample into ions. There is a wide variety of ionization techniques, depending on the phase (solid, liquid, gas) of the sample and the efficiency of various ionization mechanisms for the unknown species. An extraction system removes ions from the sample, which are then trajected through the mass analyzer and onto the *detector*. The differences in masses of the fragments allows the mass analyzer to sort the ions by their mass-to-charge ratio. The detector measures the value of an indicator quantity and thus provides data for calculating the abundances of each ion present. Some detectors also give spatial information, e.g., a multichannel plate.

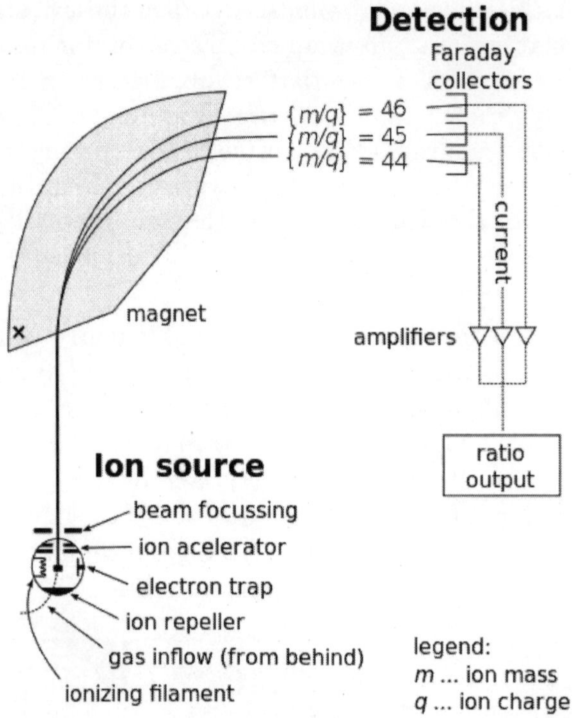

Figure: *Schematics of a simple mass spectrometre with sector type mass analyzer. This one is for the measurement of carbon dioxide isotope ratios (IRMS) as in the carbon-13 urea breath test*

Theoretical Example

The following example describes the operation of a spectrometre mass analyzer, which is of the sector type. (Other analyzer types are treated below.) Consider a sample of sodium chloride (table salt). In the ion source, the sample is vaporized (turned into gas) and ionized (transformed into electrically charged particles) into sodium (Na^+) and chloride (Cl^-) ions. Sodium atoms and ions are monoisotopic, with a mass of about 23 u. Chloride atoms and ions come in two isotopes with masses of approximately 35 u (at a natural abundance of about 75 percent) and approximately 37 u (at a natural abundance of about 25 percent). The analyzer part of the spectrometre contains electric and magnetic fields, which exert forces on ions travelling through these fields. The speed of a charged particle may be increased or decreased while passing through the electric field, and its direction may be altered by the magnetic field. The magnitude of the deflection of the moving ion's trajectory depends on its mass-to-charge ratio. Lighter ions get deflected by the magnetic force more than heavier ions (based on Newton's second law of motion, $F = ma$). The streams of sorted ions pass from the analyzer to the detector, which records the relative abundance of each ion type. This information is used to determine the chemical element composition of the original sample (i.e. that both sodium and chlorine are present in the sample) and the isotopic composition of its constituents (the ratio of ^{35}Cl to ^{37}Cl).

Creating Ions

The ion source is the part of the mass spectrometre that ionizes the material under analysis (the analyte). The ions are then transported by magnetic or electric fields to the mass analyzer.

Techniques for ionization have been key to determining what types of samples can be analyzed by mass spectrometry. Electron ionization and chemical ionization are used for gases and vapors. In chemical ionization sources, the analyte is ionized by chemical ion-molecule reactions during collisions in the source. Two techniques often used with liquid and solid biological samples include electrospray ionization (invented by John Fenn) and matrix-assisted laser desorption/ionization (MALDI, initially developed as a similar technique "Soft Laser Desorption (SLD)" by K. Tanaka for which a Nobel Prize was awarded and as MALDI by M. Karas and F. Hillenkamp).

Inductively Coupled Plasma

Inductively coupled plasma (ICP) sources are used primarily for cation analysis of a wide array of sample types. In this type of Ion Source Technology, a 'flame' of plasma that is electrically neutral overall, but that has had a substantial fraction of its atoms ionized by high temperature, is used to atomize introduced sample molecules and to further strip the outer electrons from those atoms. The plasma is usually generated from argon gas, since the first ionization energy of argon atoms is higher than the first of any other elements except He, O, F and Ne, but lower than the second ionization energy of all except the most electropositive metals. The heating is achieved by a radio-frequency current passed through a coil surrounding the plasma.

Other Ionization Techniques

Others include glow discharge, field desorption (FD), fast atom bombardment (FAB), thermospray, desorption/ionization on silicon (DIOS), Direct Analysis in Real Time (DART), atmospheric pressure chemical ionization (APCI), secondary ion mass spectrometry (SIMS), spark ionization and thermal ionization (TIMS). Ion attachment ionization is an ionization technique that allows for fragmentation free analysis.

Hard Ionisation and Soft Ionization

In mass spectrometry (MS), ionisation (or ionization) refers to the production of gas phase ions suitable for resolution in the mass analyser or mass filter. Ionisation occurs in the instrument ion source. There are a plethora of ion sources available, each has advantages and disadvantages for particular applications. For example, Electron Impact (EI) ionisation gives a high degree of fragmentation, yielding highly detailed mass spectra which when skilfully analysed can provide important information for structural eluciation/ characterisation and facilitate identification of unknown compounds by comparison to mass spectral libraries obtained under identical operating conditions. However, EI is not suitable for coupling to HPLC, i.e. LC-MS, since at atmospheric pressure, the filaments used to generate electrons burn out rapidly. Thus EI is coupled predominantly with GC, i.e. GC-MS, where the entire system is under high vacuum.

Ionisation techniques can be described as belonging to one of the following categories:

- Hard ionisation
- Soft ionisation

Hard Ionization

Hard ionisation techniques are processes which impart high quantities of residual energy in the subject molecule invoking large degrees of fragmentation (i.e. the systematic rupturing of bonds acts to remove the excess energy, restoring stability to the resulting ion). Resultant ions tend to have m/z lower than the molecular mass (other than in the case of proton transfer and not including isotope peaks). The most common example of hard ionisation is Electron Impact Ionisation (EI).

Soft Ionization

Soft ionisation refers to the processes which impart very little residual energy onto the subject molecule and as such result in very little fragmentation. Examples include:

- Fast atom bombardment (FAB)
- Chemical ionisation (CI)
- Atmospheric Pressure Chemical Ionisation (APCI)
- Electrospray Ionisation (ESI)
- Matrix Assisted Laser Desorption Ionisation (MALDI)

Mass Selection

Mass analyzers separate the ions according to their mass-to-charge ratio. The following two laws govern the dynamics of charged particles in electric and magnetic fields in vacuum:

$F = Q(E + v \times B)$ (Lorentz force law);

$F = ma$ (Newton's second law of motion in non-relativistic case, i.e. valid only at ion velocity much lower than the speed of light).

Here F is the force applied to the ion, m is the mass of the ion, a is the acceleration, Q is the ion charge, E is the electric field, and v × B is the vector cross product of the ion velocity and the magnetic field

Equating the above expressions for the force applied to the ion yields:

$$(m / Q)\mathrm{a} = \mathrm{E} + \mathrm{v} \times \mathrm{B}.$$

This differential equation is the classic equation of motion for charged particles. Together with the particle's initial conditions, it completely determines the particle's motion in space and time in terms of m/Q. Thus mass spectrometres could be thought of as "mass-to-charge spectrometres". When presenting data, it is common to use the (officially) dimensionless m/z, where z is the number of elementary charges (e) on the ion (z=Q/e). This quantity, although it is informally called the mass-to-charge ratio, more accurately speaking represents the ratio of the mass number and the charge number, z.

There are many types of mass analyzers, using either static or dynamic fields, and magnetic or electric fields, but all operate according to the above differential equation. Each analyzer type has its strengths and weaknesses. Many mass spectrometres use two or more mass analyzers for tandem mass spectrometry (MS/MS). In addition to the more common mass analyzers listed below, there are others designed for special situations.

There are several important analyser characteristics. The mass resolving power is the measure of the ability to distinguish two peaks of slightly different m/z. The mass accuracy is the ratio of the m/z measurement error to the true m/z. Mass accuracy is usually measured in ppm or milli mass units. The mass range is the range of m/z amenable to analysis by a given analyzer. The linear dynamic range is the range over which ion signal is linear with analyte concentration. Speed refers to the time frame of the experiment and ultimately is used to determine the number of spectra per unit time that can be generated.

Sector Instruments

A sector field mass analyzer uses an electric and/or magnetic field to affect the path and/or velocity of the charged particles in some way. As shown above, sector instruments bend the trajectories of the ions as they pass through the mass analyzer, according to their mass-to-charge ratios, deflecting the more charged and faster-moving, lighter ions more. The analyzer can be used to select a narrow range of m/z or to scan through a range of m/z to catalog the ions present.

Time-of-Flight

The time-of-flight (TOF) analyzer uses an electric field to accelerate the ions through the same potential, and then measures the time they take to reach the detector. If the particles all have the same charge, the kinetic energies will be identical, and their velocities will depend only on their masses. Lighter ions will reach the detector first.

Quadrupole Mass Filter

Quadrupole mass analyzers use oscillating electrical fields to selectively stabilize or destabilize the paths of ions passing through a radio frequency (RF) quadrupole field created between 4 parallel rods. Only the ions in a certain range of mass/charge ratio are passed through the system at any time, but changes to the potentials on the rods allow a wide range of m/z values to be swept rapidly, either continuously or in a succession of discrete hops. A quadrupole mass analyzer acts as a mass-selective filter and is closely related to the quadrupole ion trap, particularly the linear quadrupole ion trap except that it is designed to pass the untrapped ions rather than collect the trapped ones, and is for that reason referred to as a transmission quadrupole. A common variation of the transmission quadrupole is the triple quadrupole mass spectrometre. The "triple quad" has three consecutive quadrupole stages, the first acting as a mass filter to transmit a particular incoming ion to the second quadrupole, a collision chamber, wherein that ion can be broken into fragments. The third quadrupole also acts as a mass filter, to transmit a particular fragment ion to the detector. If a quadrupole is made to rapidly and repetitively cycle through a range of mass filter settings, full spectra can be reported. Likewise, a triple quad can be made to perform various scan types characteristic of tandem mass spectrometry.

Ion Traps

Three-Dimensional Quadrupole ion Trap: The quadrupole ion trap works on the same physical principles as the quadrupole mass analyzer, but the ions are trapped and sequentially ejected. Ions are trapped in a mainly quadrupole RF field, in a space defined by a ring electrode (usually connected to the main RF potential) between two endcap electrodes (typically connected to DC or auxiliary AC potentials). The sample is ionized either internally (e.g. with an electron or laser beam), or externally, in which case the ions are often introduced through an aperture in an endcap electrode.

There are many mass/charge separation and isolation methods but the most commonly used is the mass instability mode in which the RF potential is ramped so that the orbit of ions with a mass $a > b$ are stable while ions with mass b become unstable and are ejected on the z-axis onto a detector. There are also non-destructive analysis methods.

Ions may also be ejected by the resonance excitation method, whereby a supplemental oscillatory excitation voltage is applied to the endcap electrodes, and the trapping voltage amplitude and/or excitation voltage frequency is varied to bring ions into a resonance condition in order of their mass/charge ratio.

The cylindrical ion trap mass spectrometre is a derivative of the quadrupole ion trap mass spectrometre.

Linear Quadrupole ion Trap

A linear quadrupole ion trap is similar to a quadrupole ion trap, but it traps ions in a two dimensional quadrupole field, instead of a three-dimensional quadrupole field as in a 3D quadrupole ion trap. Thermo Fisher's LTQ ("linear trap quadrupole") is an example of the linear ion trap.

A toroidal ion trap can be visualized as a linear quadrupole curved around and connected at the ends or as a cross section of a 3D ion trap rotated on edge to form the toroid, donut shaped trap.

The trap can store large volumes of ions by distributing them throughout the ring-like trap structure. This toroidal shaped trap is a configuration that allows the increased miniaturization of an ion trap mass analyzer. Additionally all ions are stored in the same trapping field and ejected together simplifying detection that can be complicated with array configurations due to variations in detector alignment and machining of the arrays.

Orbitrap

An orbitrap is a type of mass analyzer invented by Alexander Makarov. It consists of an outer barrel-like electrode and a coaxial inner spindle-like electrode that form an electrostatic field with quadro-logarithmic potential distribution. Image current from dynamically trapped ions is detected, digitized and converted using Fourier transform into frequency and then mass spectra.

Principle of Operation

Trapping

In an orbitrap, ions are trapped because their electrostatic attraction to the inner electrode is balanced by centrifugal forces. Thus, ions cycle around the inner electrode on elliptical trajectories.

In addition, the ions also move back and forth along the axis of the central electrode thus their trajectories in space look like helices. Due to properties of quadro-logarithmic potential, axial motion is harmonic, i.e. it is completely independent not only of motion around the inner electrode but also of all initial parameters of ions except their mass-to-charge ratios m/q. Its angular frequency is: ω="(k/(m/q)) where k is the force constant of the potential, similar to the Spring Constant.

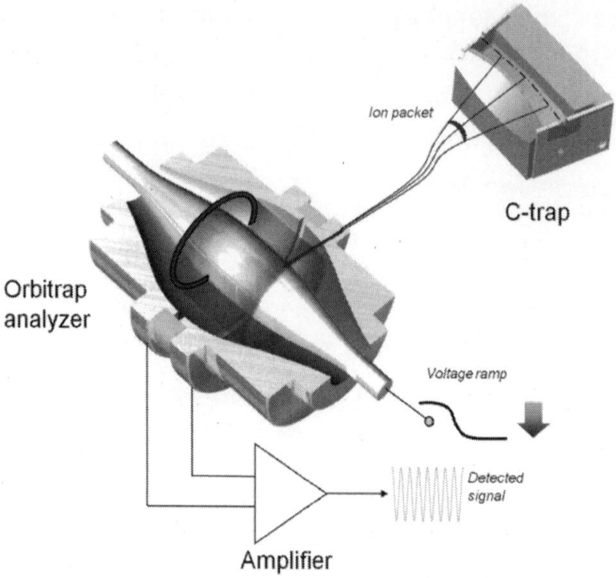

Figure: *Cross-section of the C-trap and Orbitrap analyzer (ion optics and differential pumping not shown). Ion packet enters the analyzer during the voltage ramp and form rings that induce current detected by the amplifier. Reprinted with permission of Thermo Fisher Scientific, CCL2012*

Injection

In order to inject ions from an external ion source, the field between electrodes is first reduced. As ion packets are injected tangentially into the field, the electric field is increased by ramping voltage on the inner electrode. Ions get squeezed towards the inner electrode until they reach the desired orbit inside the trap. At that moment ramping is stopped, field becomes static and detection could start. Each packet contain a multitude of ions of different velocities spread over a certain volume, therefore ions will move with different rotational frequencies but with the same axial frequency. It means that ions of a specific mass-to-charge ratio spread into rings which oscillate along the inner spindle.

Excitation

In principle, coherent axial oscillations of ion rings could be excited by applying RF waveforms to the outer electrode as demonstrated in and references therein. However, if packets are injected away from the minimum of axial potential (which corresponds to the thickest part of either electrode), this automatically initiates their axial oscillations eliminating need for any additional excitation. Furthermore, absence of excitations allows to start detection process as soon as detection electronics recovers from the voltage ramp needed for ion injection.

Detection

Axial oscillations of ion rings are detected by their image current induced on the outer electrode which is split in two symmetrical pick-up sensors connected to a differential amplifier. By processing data in a manner similar to that used in Fourier transform ion cyclotron resonance mass spectrometry (FTICR-MS), the trap can be used as a mass analyzer. Like in FTICR-MS, all ions are detected simultaneously over some given period of time and resolution can be improved by increasing the strength of the field or by increasing the detection duration. Orbitrap differs from FTICR-MS by absence of magnetic field and hence significantly slower decrease of resolving power with increasing m/z.

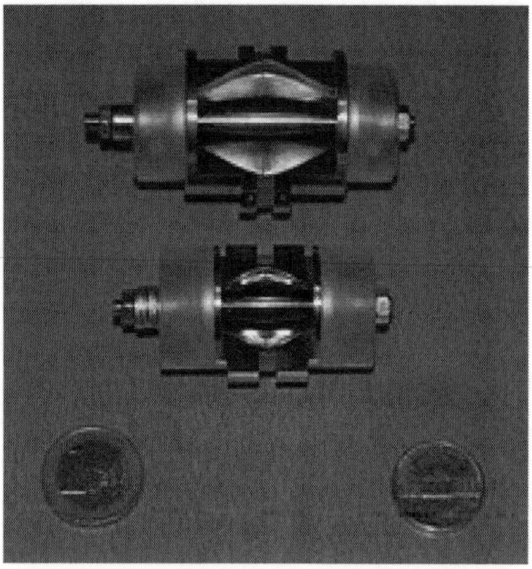

Figure: *Cut-outs of a standard (top) and a high-field (bottom) Orbitrap analyzer. Reprinted with permission of Thermo Fisher Scientific, CCL2012.*

Orbitrap Mass Spectrometry

Direct External Injection

Proof-of-principle of the technology was carried out with direct injection of ions from an external laser desorption and ionization ion source. This method of injection works well with pulsed sources like MALDI but cannot be interfaced to continuous ion sources like electrospray.

External Injection from a Trapping Device

All commercial Orbitrap mass spectrometres utilise a curved linear trap for ion injection (so-called C-trap) as shown in Figure earlier. By rapidly ramping down trapping RF voltages and applying DC gradients across the C-trap, ions could be bunched into short packets similar to those from laser ion source. The C-trap is tightly integrated with the analyzer, injection optics and differential pumping.

Variants of the Orbitrap Analyzer

Currently the Orbitrap analyzer exists in two variants: standard trap and compact high-field trap. In practical traps, the outer electrode is sustained at virtual ground and voltage of 3.5 or 5 kV is applied to the inner electrode only. As the result, resolving power at m/z 400 and 768 ms detection time could range from 60,000 for a standard trap at 3.5 kV to 280,000 for high field trap at 5kV and with enhanced FT processing. Like in FTICR-MS the orbitrap resolving power is proportional to the number of harmonic oscillations of the ions; as a result, the resolving power is inversely proportional to the square root of m/z and proportional to acquisition time. For example, the values above would double for m/z 100 and halve for m/z 1600. For shortest transient of 96 ms these values would be reduced by 8 times while resolving power in excess of 1,000,000 was demonstrated in 3 sec transients.

Practical Orbitrap Mass Spectrometres

Orbitrap analyzer is known to be interfaced to a linear ion trap (LTQ Orbitrap family of instruments), quadrupole mass filter (Q Exactive family) or directly to an ion source (Exactive instrument, all marketed by Thermo Fisher Scientific). In addition, higher-energy collision cell could be appended to the C-trap, with further addition of electron-transfer dissociation on its back. Most of these instruments have atmospheric pressure ion sources though intermediate-pressure MALDI source is also used (MALDI LTQ Orbitrap). All these instruments provide a high mass accuracy (<2-3 ppm with external calibrant and <1-2 ppm with internal), a high resolving power (up to 240,000 at m/z 400), a high dynamic range and high sensitivity.

Applications of Orbitrap Mass Spectrometry

Orbitrap-based mass spectrometres became a workhorse in proteomics and are also widely used in all major applications of life science mass spectrometry such as metabolism, metabolomics, environmental, food and safety analysis. Most of them are interfaced to liquid chromatography separations, though they are also used with gas chromatography and ambient ionization methods.

Fourier Transform Ion Cyclotron Resonance

Fourier transform mass spectrometry (FTMS), or more precisely Fourier transform ion cyclotron resonance MS, measures mass by detecting the image current produced by ions cyclotroning in the presence of a magnetic field. Instead of measuring the deflection of ions with a detector such as an electron multiplier, the ions are injected into a Penning trap (a static electric/magnetic ion trap) where they effectively form part of a circuit. Detectors at fixed positions in space measure the electrical signal of ions which pass near them over time, producing a periodic signal. Since the frequency of an ion's cycling is determined by its mass to charge ratio, this can be deconvoluted by performing a Fourier transform on the signal. FTMS has the advantage of high sensitivity (since each ion is "counted" more than once) and much higher resolution and thus precision.

Ion cyclotron resonance (ICR) is an older mass analysis technique similar to FTMS except that ions are detected with a traditional detector. Ions trapped in a Penning trap are excited by an RF electric field until they impact the wall of the trap, where the detector is located. Ions of different mass are resolved according to impact time.

Detectors

Figure: *A continuous dynode particle multiplier detector.*

The final element of the mass spectrometre is the detector. The detector records either the charge induced or the current produced when an ion passes by or hits a surface. In a scanning instrument, the signal produced in the detector during the course of the scan versus where the instrument is in the scan (at what m/Q) will produce a mass spectrum, a record of ions as a function of m/Q.

Typically, some type of electron multiplier is used, though other detectors including Faraday cups and ion-to-photon detectors are also used. Because the number of ions leaving the mass analyzer at a particular instant is typically quite small, considerable amplification is often necessary to get a signal. Microchannel plate detectors are commonly used in modern commercial instruments. In FTMS and Orbitraps, the detector consists of a pair of metal surfaces within the mass analyzer/ion trap region which the ions only pass near as they oscillate. No direct current is produced, only a weak AC image current is produced in a circuit between the electrodes. Other inductive detectors have also been used.

Tandem Mass Spectrometry

A tandem mass spectrometre is one capable of multiple rounds of mass spectrometry, usually separated by some form of molecule fragmentation. For example, one mass analyzer can isolate one peptide from many entering a mass spectrometre. A second mass analyzer then stabilizes the peptide ions while they collide with a gas, causing them to fragment by collision-induced dissociation (CID). A third mass analyzer then sorts the fragments produced from the peptides. Tandem MS can also be done in a single mass analyzer over time, as in a quadrupole ion trap. There are various methods for fragmenting molecules for tandem MS, including collision-induced dissociation (CID), electron capture dissociation (ECD), electron transfer dissociation (ETD), infrared multiphoton dissociation (IRMPD), blackbody infrared radiative dissociation (BIRD), electron-detachment dissociation (EDD) and surface-induced dissociation (SID). An important application using tandem mass spectrometry is in protein identification.

Tandem mass spectrometry enables a variety of experimental sequences. Many commercial mass spectrometres are designed to expedite the execution of such routine sequences as selected reaction monitoring (SRM) and precursor ion scanning. In SRM, the first analyzer allows only a single mass through and the second analyzer monitors for multiple user-defined fragment ions. SRM is most often used with scanning instruments where the second mass analysis event is duty cycle limited. These experiments are used to increase specificity of detection of known molecules, notably in pharmacokinetic studies. Precursor ion scanning refers to monitoring for a specific loss from the precursor ion. The first and second mass analyzers scan across the spectrum as partitioned by a user-defined m/z value. This experiment is used to detect specific motifs within unknown molecules.

Another type of tandem mass spectrometry used for radiocarbon dating is accelerator mass spectrometry (AMS), which uses very high voltages, usually in the mega-volt range, to accelerate negative ions into a type of tandem mass spectrometre.

Common Mass Spectrometre Configurations and Techniques

When a specific configuration of source, analyzer, and detector becomes conventional in practice, often a compound acronym arises to designate it, and the compound acronym may be better known among nonspectrometrists than the component acronyms. The epitome of this is MALDI-TOF, which simply refers to combining a matrix-assisted laser desorption/ionization source with a time-of-flight mass analyzer. The MALDI-TOF moniker is more widely recognised by the non-mass spectrometrists than MALDI or TOF individually. Other examples include inductively coupled plasma-mass spectrometry (ICP-

MS), accelerator mass spectrometry (AMS), thermal ionization-mass spectrometry (TIMS) and spark source mass spectrometry (SSMS). Sometimes the use of the generic "MS" actually connotes a very specific mass analyzer and detection system, as is the case with AMS, which is always sector based.

Certain applications of mass spectrometry have developed monikers that although strictly speaking would seem to refer to a broad application, in practice have come instead to connote a specific or a limited number of instrument configurations. An example of this is isotope ratio mass spectrometry (IRMS), which refers in practice to the use of a limited number of sector based mass analyzers; this name is used to refer to both the application and the instrument used for the application.

Chromatographic Techniques Combined with Mass Spectrometry

An important enhancement to the mass resolving and mass determining capabilities of mass spectrometry is using it in tandem with chromatographic separation techniques.

Gas Chromatography

A common combination is gas chromatography-mass spectrometry (GC/MS or GC-MS). In this technique, a gas chromatograph is used to separate different compounds. This stream of separated compounds is fed online into the ion source, a metallic filament to which voltage is applied. This filament emits electrons which ionize the compounds. The ions can then further fragment, yielding predictable patterns. Intact ions and fragments pass into the mass spectrometre's analyzer and are eventually detected.

Liquid Chromatography

Similar to gas chromatography MS (GC/MS), liquid chromatography mass spectrometry (LC/MS or LC-MS) separates compounds chromatographically before they are introduced to the ion source and mass spectrometre. It differs from GC/MS in that the mobile phase is liquid, usually a mixture of water and organic solvents, instead of gas. Most commonly, an electrospray ionization source is used in LC/MS. Other popular and commercially available LC/MS ion sources are atmospheric pressure chemical ionization and atmospheric pressure photoionization. There are also some newly developed ionization techniques like laser spray.

Ion Mobility

Ion mobility spectrometry/mass spectrometry (IMS/MS or IMMS) is a technique where ions are first separated by drift time through some neutral gas under an applied electrical potential gradient before being introduced into a mass spectrometre. Drift time is a measure of the radius relative to the charge of the ion. The duty cycle of IMS (the time over which the experiment takes place) is longer than most mass spectrometric techniques, such that the mass spectrometre can sample along the course of the IMS separation. This produces data about the IMS separation and the mass-to-charge ratio of the ions in a manner similar to LC/MS.

The duty cycle of IMS is short relative to liquid chromatography or gas chromatography separations and can thus be coupled to such techniques, producing triple modalities such as LC/IMS/MS.

Data and Analysis

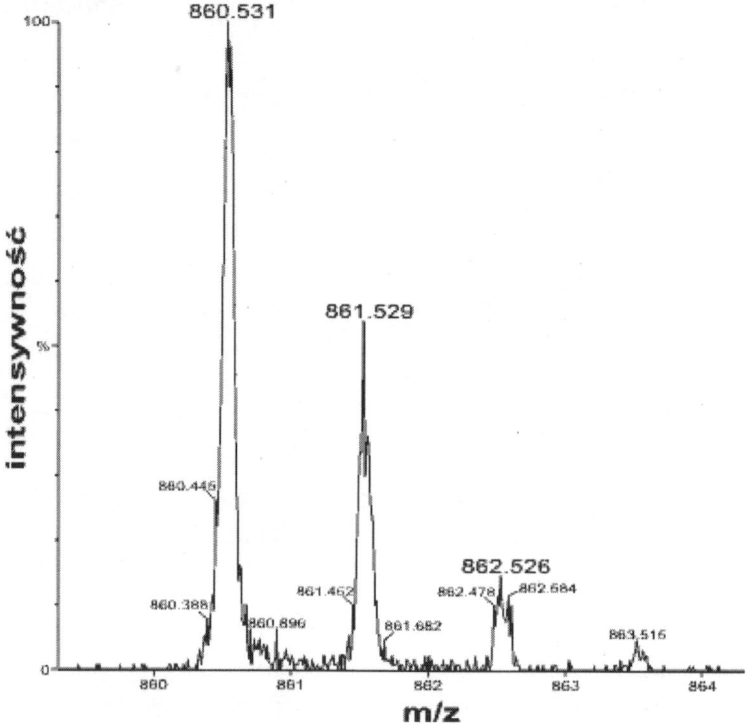

Figure: *Mass spectrum of a peptide showing the isotopic distribution*

Data representations

Mass spectrometry produces various types of data. The most common data representation is the mass spectrum. Certain types of mass spectrometry data are best represented as a mass chromatogram. Types of chromatograms include selected ion monitoring (SIM), total ion current (TIC), and selected reaction monitoring (SRM), among many others.

Other types of mass spectrometry data are well represented as a three-dimensional contour map. In this form, the mass-to-charge, m/z is on the x-axis, intensity the y-axis, and an additional experimental parameter, such as time, is recorded on the z-axis.

Data Analysis

Basics: Mass spectrometry data analysis is specific to the type of experiment producing the data. General subdivisions of data are fundamental to understanding any data. Many mass spectrometres work in either *negative ion mode* or *positive ion mode.* It is very important to know whether the observed ions are negatively or positively charged. This is often important in determining the neutral mass but it also indicates something about the nature of the molecules. Different types of ion source result in different arrays of fragments produced from the original molecules. An electron ionization source produces many fragments and mostly single-charged (1-) radicals (odd number of electrons), whereas an electrospray source usually produces non-radical quasimolecular ions that are frequently multiply charged. Tandem mass spectrometry purposely produces fragment ions post-source and can drastically change the sort of data achieved by an experiment.

Knowledge of the origin of a sample can provide insight into the component molecules of the sample and their fragmentations. A sample from a synthesis/ manufacturing process will probably contain impurities chemically related to the target component. A crudely prepared biological sample will probably contain a certain amount of salt, which may form adducts with the analyte molecules in certain analyses. Results can also depend heavily on sample preparation and how it was run/introduced. An important example is the issue of which matrix is used for MALDI spotting, since much of the energetics of the desorption/ionization event is controlled by the matrix rather than the laser power. Sometimes samples are spiked with sodium or another ion-carrying species to produce adducts rather than a protonated species.

Mass spectrometry can measure molar mass, molecular structure, and sample purity. Each of these questions requires a different experimental procedure; therefore, adequate definition of the experimental goal is a prerequisite for collecting the proper data and successfully interpreting it.

Interpretation of Mass Spectra

Since the precise structure or peptide sequence of a molecule is deciphered through the set of fragment masses, the interpretation of mass spectra requires combined use of various techniques. Usually the first strategy for identifying an unknown compound is to compare its experimental mass spectrum against a library of mass spectra. If no matches result from the search, then manual interpretation or software assisted interpretation of mass spectra must be performed. Computer simulation of ionization and fragmentation processes occurring in mass spectrometre is the primary tool for assigning structure or peptide sequence to a molecule. An *a priori* structural information is fragmented *in silico* and the resulting pattern is compared with observed spectrum. Such simulation is often supported by a fragmentation library that contains published patterns of known decomposition reactions. Software taking advantage of this idea has been developed for both small molecules and proteins.

Analysis of mass spectra can also be spectra with accurate mass. A mass-to-charge ratio value (m/z) with only integer precision can represent an immense number of theoretically possible ion structures; however, more precise mass figures significantly reduce the number of candidate molecular formulas. A computer algorithm called formula generator calculates all molecular formulas that theoretically fit a given mass with specified tolerance.

A recent technique for structure elucidation in mass spectrometry, called precursor ion fingerprinting, identifies individual pieces of structural information by conducting a search of the tandem spectra of the molecule under investigation against a library of the production spectra of structurally characterized precursor ions.

Applications

Mass spectrometry has both qualitative and quantitative uses. These include identifying unknown compounds, determining the isotopic composition of elements in a molecule, and determining the structure of a compound by observing its fragmentation. Other uses include quantifying the amount of a compound in a sample or studying the fundamentals of gas phase ion chemistry (the chemistry of ions and neutrals in a vacuum). MS is now in very common use in analytical laboratories that study physical, chemical, or biological properties of a great variety of compounds.

As an analytical technique it possesses distinct advantages such as: 1. Increased sensitivity over most other analytical techniques because the analyzer, as a mass-charge filter, reduces background interference 2. Excellent specificity from characteristic fragmentation patterns to identify unknowns or confirm the presence of suspected compounds. 3. Information about molecular weight. 4. Information about the isotopic abundance of elements. 5. Temporally resolved chemical data.

A few of the disadvantages of the method is that often fails to distinguish between optical and geometrical isomers and the positions of substituent in o-, m- and p- positions in an aromatic ring. Also, its scope is limited in identifying hydrocarbons that produce similar fragmented ions.

Isotope Ratio MS: Isotope Dating and Tracking

Mass spectrometry is also used to determine the isotopic composition of elements within a sample. Differences in mass among isotopes of an element are very small, and the less abundant isotopes of an element are typically very rare, so a very sensitive instrument is required. These instruments, sometimes referred to as isotope ratio mass spectrometres (IR-MS), usually use a single magnet to bend a beam of ionized particles towards a series of Faraday cups which convert particle impacts to electric current.

A fast on-line analysis of deuterium content of water can be done using Flowing afterglow mass spectrometry, FA-MS. Probably the most sensitive and accurate mass spectrometre for this purpose is the accelerator mass spectrometre (AMS). Isotope ratios are important markers of a variety of processes. Some isotope ratios are used to determine the age of materials for example as in carbon dating. Labeling with stable isotopes is also used for protein quantification.

Trace Gas Analysis

Several techniques use ions created in a dedicated ion source injected into a flow tube or a drift tube: selected ion flow tube (SIFT-MS), and proton transfer reaction (PTR-MS), are variants of chemical ionization dedicated for trace gas analysis of air, breath or liquid headspace using well defined reaction time allowing calculations of analyte concentrations from the known reaction kinetics without the need for internal standard or calibration.

Atom Probe

An atom probe is an instrument that combines time-of-flight mass spectrometry and field ion microscopy (FIM) to map the location of individual atoms.

Pharmacokinetics

Pharmacokinetics is often studied using mass spectrometry because of the complex nature of the matrix (often blood or urine) and the need for high sensitivity to observe low dose and long time point data. The most common instrumentation used in this application is LC-MS with a triple quadrupole mass spectrometre. Tandem mass spectrometry is usually employed for added specificity. Standard curves and internal standards are used for quantitation of usually a single pharmaceutical in the samples. The samples represent different time points as a pharmaceutical is administered and then metabolized or cleared from the body. Blank or t=0 samples taken before administration are important in determining background and ensuring data integrity with such complex sample matrices. Much attention is paid to the linearity of the standard curve; however it is not uncommon to use curve fitting with more complex functions such as quadratics since the response of most mass spectrometres is less than linear across large concentration ranges.

There is currently considerable interest in the use of very high sensitivity mass spectrometry for microdosing studies, which are seen as a promising alternative to animal experimentation.

Protein Characterization

Mass spectrometry is an important method for the characterization and sequencing of proteins. The two primary methods for ionization of whole proteins are electrospray ionization (ESI) and matrix-assisted laser desorption/ ionization (MALDI). In keeping with the performance and mass range of available mass spectrometres, two approaches are used for characterizing proteins. In the first, intact proteins are ionized by either of the two techniques described above, and then introduced to a mass analyzer. This approach is referred to as "top-down" strategy of protein analysis. In the second, proteins are enzymatically digested into smaller peptides using proteases such as trypsin or pepsin, either in solution or in gel after electrophoretic separation. Other proteolytic agents are also used. The collection of peptide products are then introduced to the mass analyzer. When the characteristic pattern of peptides is used for the identification of the protein the method is called peptide mass fingerprinting (PMF), if the identification is performed using the sequence data determined in tandem MS analysis it is called de novo sequencing. These procedures of protein analysis are also referred to as the "bottom-up" approach.

Glycan Analysis

Mass spectrometry (MS), with its low sample requirement and high sensitivity, has been predominantly used in glycobiology for characterization and elucidation of glycan structures. Mass spectrometry provides a complementary method to HPLC for the analysis of glycans. Intact glycans may be detected directly as singly charged ions by matrix-assisted laser desorption/ionization mass spectrometry (MALDI-MS) or, following permethylation or peracetylation, by fast atom bombardment mass spectrometry (FAB-MS). Electrospray ionization mass spectrometry (ESI-MS) also gives good signals for the smaller glycans. Various free and commercial software are now available which interpret MS data and aid in Glycan structure characterization.

Space Exploration

As a standard method for analysis, mass spectrometres have reached other planets and moons. Two were taken to Mars by the Viking program. In early 2005 the Cassini–Huygens mission delivered a specialised GC-MS instrument aboard the Huygens probe through the atmosphere of Titan, the largest moon of the planet Saturn. This instrument analyzed atmospheric samples along its descent trajectory and was able to vaporize and analyze samples of Titan's frozen, hydrocarbon covered surface once the probe had landed. These measurements compare the abundance of isotope(s) of each particle comparatively to earth's natural abundance. Also on board the Cassini–Huygens spacecraft is an ion and neutral mass spectrometre which has been taking measurements of Titan's atmospheric composition as well as the composition of Enceladus' plumes. A Thermal and Evolved Gas Analyzer mass spectrometre was carried by the Mars Phoenix Lander launched in 2007. Mass spectrometres are also widely used in space missions to measure the composition of plasmas. For example, the Cassini spacecraft carries the Cassini Plasma Spectrometre (CAPS), which measures the mass of ions in Saturn's magnetosphere.

Respired Gas Monitor

Mass spectrometres were used in hospitals for respiratory gas analysis beginning around 1975 through the end of the century. Some are probably still in use but none are currently being manufactured. Found mostly in the operating room, they were a part of a complex system, in which respired gas samples from patients undergoing anesthesia were drawn into the instrument through a valve mechanism designed to sequentially connect up to 32 rooms to the mass spectrometre. A computer directed all operations of the system. The data collected from the mass spectrometre was delivered to the individual rooms for the anesthesiologist to use.

The uniqueness of this magnetic sector mass spectrometre may have been the fact that a plane of detectors, each purposely positioned to collect all of the ion species expected to be in the samples, allowed the instrument to simultaneously report all of the gases respired by the patient. Although the mass range was limited to slightly over 120 u, fragmentation of some of the heavier molecules negated the need for a higher detection limit.

Electrochemical pH Sensors

A pH metre is an electronic device used for measuring the pH (acidity or alkalinity) of a liquid (though special probes are sometimes used to measure the pH of semi-solid substances). A typical pH metre consists of a special measuring probe (a glass electrode) connected to an electronic metre that measures and displays the pH reading. The probe is a key part of a pH metre, it is a rod like structure usually made up of glass. At the bottom of the probe there is a bulb, the bulb is a sensitive part of a probe that contains the sensor. Never touch the bulb by hand and clean it with the help of an absorbent tissue paper with very soft hands, being careful not to rub the tissue against the glass bulb in order to avoid creating static. To measure the pH of a solution, the probe is dipped into the solution. The probe is fitted in an arm known as the probe arm.

Calibration and Use

For very precise work the pH metre should be calibrated before each measurement. For normal use calibration should be performed at the beginning of each day. The reason for this is that the glass electrode does not give a reproducible e.m.f. over longer periods of time. Calibration should be performed with at least two standard buffer solutions that span the range of pH values to be measured. For general purposes buffers at pH 4.01 and pH 10.00 are acceptable. The pH metre has one control (calibrate) to set the metre reading equal to the value of the first standard buffer and a second control (slope) which is used to adjust the metre reading to the value of the second buffer. A third control allows the temperature to be set. Standard buffer sachets, which can be obtained from a variety of suppliers, usually state how the buffer value changes with temperature. For more precise measurements, a three buffer solution calibration is preferred. As pH 7 is essentially, a "zero point" calibration (akin to zeroing or taring a scale or balance), calibrating at pH 7 first, calibrating at the pH closest to the point of interest (e.g. either 4 or 10) second and checking the third point will provide a more linear accuracy to what is essentially a non-linear problem. Some metres will allow a three point calibration and that is the preferred scheme for the most accurate work. Higher quality metres will have a provision to account for temperature coefficient correction, and high-end pH probes have temperature probes built in. The calibration process correlates the voltage produced by the probe (approximately 0.06 volts per pH unit) with the pH scale. After each single measurement, the probe is rinsed with distilled water or deionized water to remove any traces of the solution being measured, blotted with a scientific wipe to absorb any remaining water which could dilute the sample and thus alter the reading, and then quickly immersed in another solution.

Storage Conditions of the Glass Probes

When not in use, the glass probe tip must be kept wet at all times to avoid the pH sensing membrane dehydration and the subsequent dysfunction of the electrode.

A glass electrode alone without combined reference electrode is typically stored immersed in an acidic solution of around pH 3.0. In an emergency, acidified tap water can be used, but distilled or deionised water must never be used for longer-term probe storage as the relatively ionless water "sucks" ions out of the probe membrane through diffusion, which degrades it.

Combined electrodes (glass membrane + reference electrode) are better stored immersed in the bridge electrolyte (often KCl 3 M) to avoid the diffusion of the electrolyte (KCl) out of the liquid junction.

Cleaning and Troubleshooting of the Glass Probes

Occasionally (about once a month), the probe may be cleaned using pH-electrode cleaning solution; generally a 0.1 M solution of hydrochloric acid (HCl) is used, having a pH of one.

Alternatively a dilute solution of ammonium fluoride (NH_4F) can be used. To avoid unexpected problems, the best practice is however to always refer to the electrode manufacturer recommendations or to a classical textbook of analytical chemistry.

Types of pH Metres

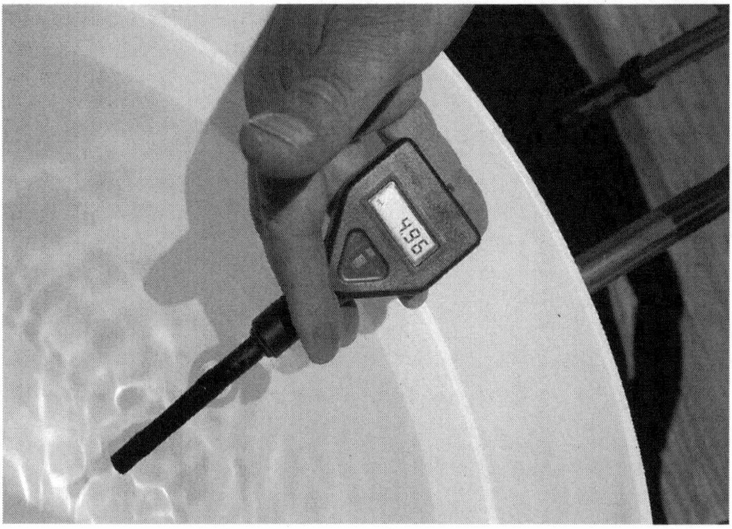

Figure: *A simple pH metre*

pH metres range from simple and inexpensive pen-like devices to complex and expensive laboratory instruments with computer interfaces and several inputs for indicator and temperature measurements to be entered to adjust for the slight variation in pH caused by temperature. Speciality metres and probes are available for use in special applications, harsh environments, etc. There are also holographic pH sensors, which allow the measurement pH colorimetrically.

History

The first commercial pH metres were built around 1936 by Radiometre in Denmark and by Arnold Orville Beckman in the United States. While Beckman was an assistant professor of chemistry at the California Institute of Technology, he was asked to devise a quick and accurate method for measuring the acidity of lemon juice for the California Fruit Growers Exchange (Sunkist). Beckman's invention helped him to launch the Beckman Instruments company (now Beckman Coulter). In 2004 the Beckman pH metre was designated an ACS National Historic Chemical Landmark in recognition of its significance as the first commercially successful electronic pH metre.

In the 1970s Jenco Electronics of Taiwan designed and manufactured the first portable digital pH metre. This metre was sold under Cole-Parmer's label.

Building a pH Metre

Because the circuitry of a basic pH metre is quite simple, it is possible to build a serviceable pH metre or pH controller with parts available at a neighbourhood electronics retailer. (pH probes, however, are not so easily acquired and must usually be ordered from a scientific instrument supplier.) For a walkthrough of how to build the simplest possible pH metre or a detailed description of how to build a pH metre/pH controller.

The application note for the LM6001 chip at the National Semiconductor web site also has a very simple demonstration circuit. Although the application note is for a speciality IC, serviceable pH metres can be built from any operational amplifier with a high input impedance, such as the common and inexpensive National Semiconductor TL082 or its equivalent.

pH Indicator

A pH indicator is a halochromic chemical compound that is added in small amounts to a solution so that the pH (acidity or basicity) of the solution can be determined visually. Hence a pH indicator is a chemical detector for hydronium ions (H_3O^+) or hydrogen ions (H^+) in the Arrhenius model. Normally, the indicator causes the colour of the solution to change depending on the pH. Indicators can also show change in other physical properties; for example, olfactory indicators show change in their odor.

At 25 °C, considered the standard temperature, the pH value of a neutral solution is 7.0. Solutions with a pH value below 7.0 are considered acidic, whereas solutions with pH value above 7.0 are basic (alkaline). As most naturally occurring organic compounds are weak protolytes, carboxylic acids and amines, pH indicators find many applications in biology and analytical chemistry. Moreover, pH indicators form one of the three main types of indicator compounds used in chemical analysis. For the quantitative analysis of metal cations, the use of complexometric indicators is preferred, whereas the third compound class, the redox indicators, are used in titrations involving a redox reaction as the basis of the analysis.

Theory

In and of themselves, pH indicators are frequently weak acids or weak bases. The general reaction scheme of a pH indicator can be formulated as follows:

$$HInd + H_2O \rightleftharpoons H_3O^+ + Ind^-$$

Here HInd stands for the acid form and Ind^- for the conjugate base of the indicator. It is the ratio of these that determines the colour of the solution and that connects the colour to the pH value. For pH indicators that are weak protolytes, we can write the Henderson-Hasselbalch equation for them:

$$pH = pK_a + \log\frac{[Ind^-]}{[HInd]}$$

The equation, derived from the acidity constant, states that when pH equals the pKa value of the indicator, both species are present in 1:1 ratio. If pH is above the pKa value, the concentration of the conjugate base is greater than the concentration of the acid, and the colour associated with the conjugate base dominates. If pH is below the pKa value, the converse is true.

Usually, the colour change is not instantaneous at the pKa value, but there is a pH range where a mixture of colours is present. This pH range varies between indicators, but as a rule of thumb, it falls between the pKa value plus or minus one. This assumes that solutions retain their colour as long as at least 10% of the other species persists. For example, if the concentration of the conjugate base is ten times greater than the concentration of the acid, their ratio is 10:1, and consequently the pH is pKa + 1. Conversely, if there is a tenfold excess of the acid with respect to the base, the ratio is 1:10 and the pH is pKa – 1.

For optimal accuracy, the colour difference between the two species should be as clear as possible, and the narrower the pH range of the colour change the better. In some indicators, such as phenolphthalein, one of the species is colourless, whereas in other indicators, such as methyl red, both species confer a colour. While pH indicators work efficiently at their designated pH range, they are usually destroyed at the extreme ends of the pH scale due to undesired side-reactions.

Application

pH indicators are frequently employed in titrations in analytical chemistry and biology to determine the extent of a chemical reaction. Because of the subjective choice (determination) of colour, pH indicators are susceptible to imprecise readings. For applications requiring precise measurement of pH, a pH metre is frequently used. Sometimes a blend of different indicators is used to achieve several smooth colour changes over a wide range of pH values. These commercial indicators (e.g., universal indicator and Hydrion papers) are used when only rough knowledge of pH is necessary.

Tabulated below are several common laboratory pH indicators. Indicators usually exhibit intermediate colours at pH values inside the listed transition range. For example, phenol red exhibits an orange colour between pH 6.8 and pH 8.4. The transition range may shift slightly depending on the concentration of the indicator in the solution and on the temperature at which it is used.

Precise pH Measurement

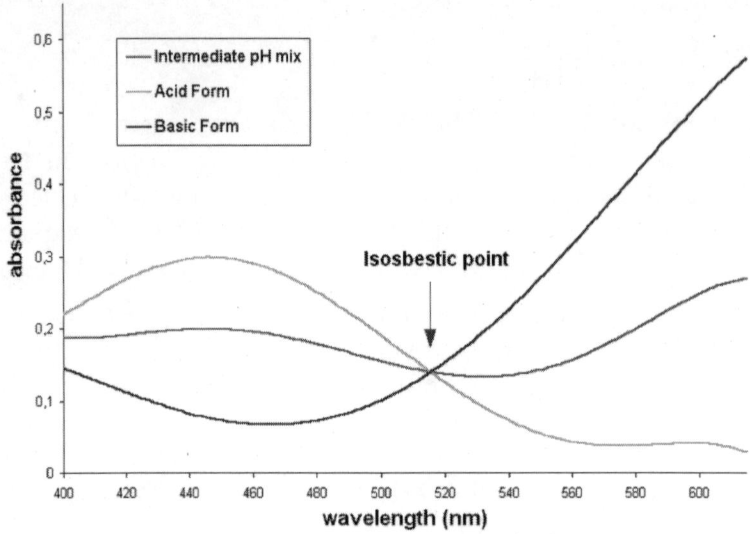

Figure: *Absorption spectra of bromocresol green at different stages of protonation.*

An indicator may be used to obtain quite precise measurements of pH by measuring absorbance quantitatively at two or more wavelengths. The principle can be illustrated by taking the indicator to be a simple acid, HA, which dissociates into H^+ and A^-.

$$HA \rightleftharpoons H^+ + A^-$$

The value of the acid dissociation constant, pK_a, must be known. The molar absorbances, ε_{HA} and ε_{A^-} of the two species HA and A^- at wavelengths λ_x and λ_y must also have been determined by previous experiment. Assuming that Beer's law is obeyed, the measured absorbances A_x and A_y at the two wavelengths are simply the sum of the absorbances due to each species.

$$A_x = [HA]\epsilon^x_{HA} + [A^-]\epsilon^x_{A^-}$$

$$A_y = [HA]\epsilon^y_{HA} + [A^-]\epsilon^y_{A^-}$$

These are two equation in the two concentrations [HA] and $[A^-]$. Once solved, the pH is obtained as

$$\mathrm{pH} = \mathrm{pK}_a + \log\frac{[\mathrm{A}^-]}{[\mathrm{HA}]}$$

If measurements are made at more than two wavelengths the concentrations [HA] and $[A^-]$ can be calculated by linear least squares. In fact a whole spectrum may be used for this purpose. The process is illustrated for the indicator bromocresol green. The observed spectrum (green) is the sum of the spectra of HA (gold) and of A- (blue), weighted for the concentration of the two species.

When a single indicator is used this method is limited to measurements in the pH range $pK_a \pm 1$, but this range can be extended by using mixtures of two or more indicators. Because indicators have intense absorption spectra the indicator concentration is relatively low so that it can usually be assumed that the indicator itself has negligible effect on pH.

Equivalence Point

In acid-base titrations, an unfitting pH indicator may induce a colour change in the indicator-containing solution before or after the actual equivalence point. As a result, different equivalence points for a solution can be concluded based on the pH indicator used. This is because the slightest colour change of the indicator-containing solution suggests the equivalence point has been reached. Therefore, the most suitable pH indicator has an effective pH range, where the change in colour is apparent, that encompasses the pH of the equivalence point of the solution being titrated.

Naturally Occurring pH Indicators

Many plants or plant parts contain chemicals from the naturally-coloured anthocyanin family of compounds. They are red in acidic solutions and blue in basic. Anthocyanins can be extracted with water or other solvents from a multitude of coloured plants or plant parts, including from leaves (red cabbage); flowers (geranium, poppy, or rose petals); berries (blueberries, blackcurrant); and stems (rhubarb). Extracting anthocyanins from household plants, especially red cabbage, to form a crude pH indicator is a popular introductory chemistry demonstration. Litmus, used by alchemists in the Middle Ages and still readily available, is a naturally occurring pH indicator made from a mixture of lichen species, particularly Roccella tinctoria. The word litmus is literally from 'coloured moss' in Old Norse. The colour changes between red in acid solutions and blue in alkalis. The term 'litmus test' has become a widely used metaphor for any test that purports to distinguish authoritatively between alternatives.

Hydrangea macrophylla flowers can change colour depending on soil acidity. In acid soils, chemical reactions occur in the soil that make aluminium available to these plants, turning the flowers blue. In alkaline soils, these reactions cannot occur and therefore aluminium is not taken up by the plant. As a result, the flowers remain pink.

Thin-Film Bulk Acoustic Resonator

A thin-film bulk acoustic resonator (FBAR or TFBAR) is a device consisting of a piezoelectric material sandwiched between two electrodes and acoustically isolated from the surrounding medium. FBAR devices using piezoelectric films with thicknesses ranging from several micrometres down to tenth of micrometres resonate in the frequency range of roughly 100 MHz to 10 GHz. Aluminium nitride and zinc oxide are two common piezoelectric materials used in FBARs.

Applications

A common application of FBARs are radio frequency (RF) filters for use in cell phones and other wireless applications. Such filters made from a network of resonators (either in half-ladder, full-ladder, lattice or stacked topologies) and are designed to remove unwanted frequencies from being transmitted in such devices, while allowing other specific frequencies to be received and transmitted. FBAR filters can also be found in duplexers. They have partially replaced an earlier technology based on surface acoustic wave (SAW) devices, due to smaller size and increased fabrication and operating efficiencies.

FBARs can be used by microwave oscillators.

FBARs can be used in sensor applications. For instance, when a FBAR device is put under mechanical pressure its resonance frequency will shift.

FBARs can also be integrated with power amplifiers or low noise amplifiers to form an integrated module solution such as PA-duplexer module, or LNA-filter module.

4

Sputtering

Sputtering is a process whereby atoms are ejected from a solid target material due to bombardment of the target by energetic particles. It only happens when the kinetic energy of the incoming particles is much higher than conventional thermal energies (k" 1 eV). This process can lead, during prolonged ion or plasma bombardment of a material, to significant erosion of materials, and can thus be harmful. On the other hand, it is commonly utilised for thin-film deposition, etching and analytical techniques

Physics of Sputtering

Physical sputtering is driven by momentum exchange between the ions and atoms in the materials, due to collisions.

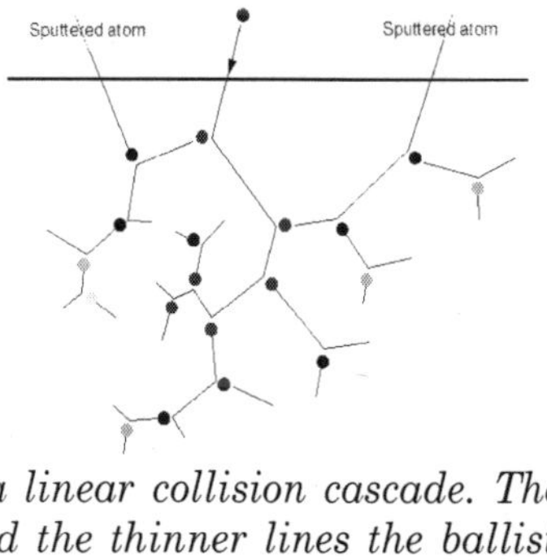

Figure: *Sputtering from a linear collision cascade. The thick line illustrates the position of the surface, and the thinner lines the ballistic movement paths of the atoms from beginning until they stop in the material. The purple circle is the incoming ion. Red, blue, green and yellow circles illustrate primary, secondary, tertiary and quaternary recoils, respectively. Two of the atoms happen to move out from the sample, i.e. be sputtered.*

The incident ions set off collision cascades in the target. When such cascades recoil and reach the target surface with an energy greater than the surface binding energy, an atom would be ejected, known as sputtering. If the target is thin on an atomic scale the collision cascade can reach the back side of the target and atoms can escape the surface binding energy "in transmission". The average number of atoms ejected from the target per incident ion is called the sputter yield and depends on the ion incident angle, the energy of the ion, the masses of the ion and target atoms, and the surface binding energy of atoms in the target. For a crystalline target the orientation of the crystal axes with respect to the target surface is relevant.

The primary particles for the sputtering process can be supplied in a number of ways, for example by a plasma, an ion source, an accelerator or by a radioactive material emitting alpha particles.

A model for describing sputtering in the cascade regime for amorphous flat targets is Thompson's analytical model. An algorithm that simulates sputtering based on a quantum mechanical treatment including electrons stripping at high energy is implemented in the program TRIM.

A different mechanism of physical sputtering is heat spike sputtering. This may occur when the solid is dense enough, and then the incoming ion heavy enough, that the collisions occur very close to each other. Then the binary collision approximation is no longer valid, but rather the collisional process should be understood as a many-body process. The dense collisions induce a heat spike (also called thermal spike), which essentially melts the crystal locally. If the molten zone is close enough to a surface, large numbers of atoms may sputter due to flow of liquid to the surface and/or microexplosions. Heat spike sputtering is most important for heavy ions (say Xe or Au or cluster ions) with energies in the keV–MeV range bombarding dense but soft metals with a low melting point (Ag, Au, Pb, etc.). The heat spike sputtering often increases nonlinearly with energy, and can for small cluster ions lead to dramatic sputtering yields per cluster of the order of 10,000. For animations of such a process see "Re: Displacement Cascade 1" in External links.

Physical sputtering has a well-defined minimum energy threshold equal to or larger than the ion energy at which the maximum energy transfer of the ion to a sample atom equals the binding energy of a surface atom. This threshold typically is somewhere in the range 10–100 eV.

Preferential sputtering can occur at the start when a multicomponent solid target is bombarded and there is no solid state diffusion. If the energy transfer is more efficient to one of the target components, and/or it is less strongly bound to the solid, it will sputter more efficiently than the other. If in an AB alloy the component A is sputtered preferentially, the surface of the solid will, during prolonged bombardment, become enriched in the B component thereby increasing the probability that B is sputtered such that the composition of the sputtered material will be AB.

Electronic Sputtering

The term electronic sputtering can mean either sputtering induced by energetic electrons (for example in a transmission electron microscope), or sputtering due to very high-energy or highly charged heavy ions that lose energy to the solid mostly by electronic stopping power, where the electronic excitations cause sputtering. Electronic sputtering produces high sputtering yields from insulators, as the electronic excitations that cause sputtering are not immediately quenched, as they would be in a conductor. One example of this is Jupiter's ice-covered moon Europa, where a MeV sulfur ion from Jupiter's magnetosphere can eject up to 10,000 H_2O molecules.

Potential Sputtering

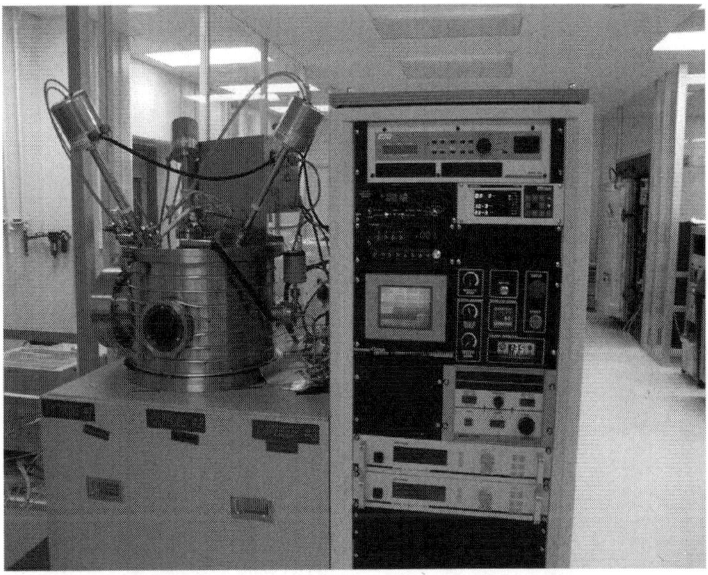

Figure: *A commercial sputtering system*

In the case of multiply charged projectile ions a particular form of electronic sputtering can take place that has been termed potential sputtering. In these cases the potential energy stored in multiply charged ions (i.e., the energy necessary to produce an ion of this charge state from its neutral atom) is liberated when the ions recombine during impact on a solid surface (formation of hollow atoms). This sputtering process is characterized by a strong dependence of the observed sputtering yields on the charge state of the impinging ion and can already take place at ion impact energies well below the physical sputtering threshold. Potential sputtering has only been observed for certain target species and requires a minimum potential energy.

Etching and Chemical Sputtering

Removing atoms by sputtering with an inert gas is called "ion milling" or "ion etching".

Sputtering can also play a role in reactive ion etching (RIE), a plasma process carried out with chemically active ions and radicals, for which the sputtering yield may be enhanced significantly compared to pure physical sputtering. Reactive ions are frequently used in secondary ion mass spectrometry (SIMS) equipment to enhance the sputter rates. The mechanisms causing the sputtering enhancement are not always well understood, but for instance the case of fluorine etching of Si has been modelled well theoretically.

Sputtering observed to occur below the threshold energy of physical sputtering is also often called chemical sputtering. The mechanisms behind such sputtering are not always well understood, and may be hard to distinguish from chemical etching. At elevated temperatures, chemical sputtering of carbon can be understood to be due to the incoming ions weakening bonds in the sample, which then desorb by thermal activation. The hydrogen-induced sputtering of carbon-based materials observed at low temperatures has been explained by H ions entering between C-C bonds and thus breaking them, a mechanism dubbed swift chemical sputtering.

Applications and Phenomena

Film Deposition

Sputter deposition is a method of depositing thin films by sputtering that involves eroding material from a "target" source onto a "substrate" e.g. a silicon wafer. Resputtering, in contrast, involves re-emission of the deposited material, e.g. SiO_2 during the deposition also by ion bombardment.

Sputtered atoms ejected into the gas phase are not in their thermodynamic equilibrium state, and tend to deposit on all surfaces in the vacuum chamber. A substrate (such as a wafer) placed in the chamber will be coated with a thin film. Sputtering usually uses an argon plasma.

Etching

In the semiconductor industry sputtering is used to etch the target. Sputter etching is chosen in cases where a high degree of etching anisotropy is needed and selectivity is not a concern. One major drawback of this technique is wafer damage.

For Analysis

Another application of sputtering is to etch away the target material. One such example occurs in secondary ion mass spectrometry (SIMS), where the target sample is sputtered at a constant rate. As the target is sputtered, the concentration and identity of sputtered atoms are measured using Mass Spectrometry. In this way the composition of the target material can be determined and even extremely low concentrations (20 μg/kg) of impurities detected. Furthermore, because the sputtering continually etches deeper into the sample, concentration profiles as a function of depth can be measured.

In Space

Sputtering is one of the forms of space weathering, a process that changes the physical and chemical properties of airless bodies, such as asteroids and the Moon. It is also one of the possible ways that Mars has lost most of its atmosphere and that Mercury continually replenishes its tenuous surface-bounded exosphere.

Ion Plating

Ion plating is a physical vapor deposition (PVD) process that is sometimes called *ion assisted deposition* (IAD) or *ion vapor deposition* (IVD) and is a version of *vacuum deposition*. Ion plating utilises concurrent or periodic bombardment of the substrate and depositing film by atomic-sized energetic particles. Bombardment prior to deposition is used to sputter clean the substrate surface. During deposition the bombardment is used to modify and control the properties of the depositing film. It is important that the bombardment be continuous between the cleaning and the deposition portions of the process to maintain an atomically clean interface.

In ion plating the energy, flux and mass of the bombarding species along with the ratio of bombarding particles to depositing particles are important processing variables. The depositing material may be vaporized either by evaporation, sputtering (bias sputtering), arc vaporization or by decomposition of a chemical vapor precursor chemical vapor deposition (CVD). The energetic particles used for bombardment are usually ions of an inert or reactive gas, or, in some cases, ions of the condensing film material ("film ions"). Ion plating can be done in a plasma environment where ions for bombardment are extracted from the plasma or it may be done in a vacuum environment where ions for bombardment are formed in a separate *ion gun*. The latter ion plating configuration is often called Ion Beam Assisted Deposition (IBAD). By using a reactive gas or vapor in the plasma, films of compound materials can be deposited.

Ion plating is used to deposit hard coatings of compound materials on tools, adherent metal coatings, optical coatings with high densities, and conformal coatings on complex surfaces.

The ion plating process was first described in the technical literature by Donald M. Mattox of Sandia National Laboratories in 1964.

Pulsed Laser Deposition (PLD)

Pulsed laser deposition (PLD) is a thin film deposition (specifically a physical vapor deposition, PVD) technique where a high-power pulsed laser beam is focused inside a vacuum chamber to strike a target of the material that is to be deposited. This material is vaporized from the target (in a plasma plume) which deposits it as a thin film on a substrate (such as a silicon wafer facing the target). This process can occur in ultra high vacuum or in the presence of a background gas, such as oxygen which is commonly used when depositing oxides to fully oxygenate the deposited films.

While the basic-setup is simple relative to many other deposition techniques, the physical phenomena of laser-target interaction and film growth are quite complex. When the laser pulse is absorbed by the target, energy is first converted to electronic excitation and then into thermal, chemical and mechanical energy resulting in evaporation, ablation, plasma formation and even exfoliation. The ejected species expand into the surrounding vacuum in the form of a plume containing many energetic species including atoms, molecules, electrons, ions, clusters, particulates and molten globules, before depositing on the typically hot substrate.

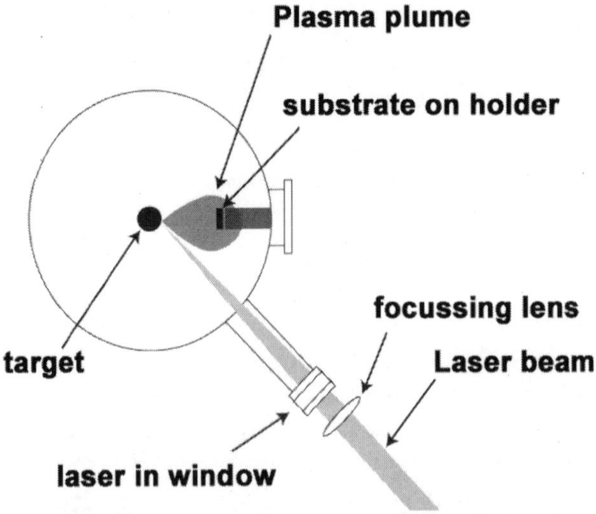

Figure: *One possible configuration of a PLD deposition chamber.*

Process

The detailed mechanisms of PLD are very complex including the ablation process of the target material by the laser irradiation, the development of a plasma plume with high energetic ions, electrons as well as neutrals and the crystalline growth of the film itself on the heated substrate. The process of PLD can generally be divided into four stages:

- Laser ablation of the target material and creation of a plasma
- Dynamic of the plasma
- Deposition of the ablation material on the substrate
- Nucleation and growth of the film on the substrate surface

Each of these steps is crucial for the crystallinity, uniformity and stoichiometry of the resulting film.

Laser Ablation of the Target Material and Creation of a Plasma

The ablation of the target material upon laser irradiation and the creation of plasma are very complex processes. The removal of atoms from the bulk material is done by vaporization of the bulk at the surface region in a state of non-equilibrium. In this the incident laser pulse penetrates into the surface of the material within the penetration depth. This dimension is dependent on the laser wavelength and the index of refraction of the target material at the applied laser wavelength and is typically in the region of 10 nm for most materials. The strong electrical field generated by the laser light is sufficiently strong to remove the electrons from the bulk material of the penetrated volume. This process occurs within 10 ps of a ns laser pulse and is caused by non-linear processes such as multiphoton ionization which are enhanced by microscopic cracks at the surface, voids, and nodules, which increase the electric field. The free electrons oscillate within the electromagnetic field of the laser light and can collide with the atoms of the bulk material thus transferring some of their energy to the lattice of the target material within the surface region. The surface of the target is then heated up and the material is vaporized.

Dynamic of the Plasma

In the second stage the material expands in a plasma parallel to the normal vector of the target surface towards the substrate due to Coulomb repulsion and recoil from the target surface. The spatial distribution of the plume is dependent on the background pressure inside the PLD chamber. The density of the plume can be described by a $\cos^n(x)$ law with a shape similar to a Gaussian curve. The dependency of the plume shape on the pressure can be described in three stages:

- The vacuum stage, where the plume is very narrow and forward directed; almost no scattering occurs with the background gases.
- The intermediate region where a splitting of the high energetic ions from the less energetic species can be observed. The time-of-flight (TOF) data can be fitted to a shock wave model; however, other models could also be possible.

- High pressure region where we find a more diffusion-like expansion of the ablated material. Naturally this scattering is also dependent on the mass of the background gas and can influence the stoichiometry of the deposited film.

The most important consequence of increasing the background pressure is the slowing down of the high energetic species in the expanding plasma plume. It has been shown that particles with kinetic energies around 50 eV can resputter the film already deposited on the substrate. This results in a lower deposition rate and can furthermore result in a change in the stoichiometry of the film.

Deposition of the Ablation Material on the Substrate

The third stage is important to determine the quality of the deposited films. The high energetic species ablated from the target are bombarding the substrate surface and may cause damage to the surface by sputtering off atoms from the surface but also by causing defect formation in the deposited film. The sputtered species from the substrate and the particles emitted from the target form a collision region, which serves as a source for condensation of particles. When the condensation rate is high enough, a thermal equilibrium can be reached and the film grows on the substrate surface at the expense of the direct flow of ablation particles and the thermal equilibrium obtained.

Nucleation and Growth of the Film on the Substrate Surface

The nucleation process and growth kinetics of the film depend on several growth parameters including:

- *Laser parameters* – several factors such as the laser fluence [Joule/cm^2], laser energy, and ionization degree of the ablated material will affect the film quality, the stoichiometry, and the deposition flux. Generally, the nucleation density increases when the deposition flux is increased.
- *Surface temperature* – The surface temperature has a large effect on the nucleation density. Generally, the nucleation density decreases as the temperature is increased.
- *Substrate surface* – The nucleation and growth can be affected by the surface preparation (such as chemical etching), the miscut of the substrate, as well as the roughness of the substrate.
- *Background pressure* – Common in oxide deposition, an oxygen background is needed to ensure stoichiometric transfer from the target to the film. If, for example, the oxygen background is too low, the film will grow off stoichiometry which will affect the nucleation density and film quality.

In PLD, a large supersaturation occurs on the substrate during the pulse duration. The pulse lasts around 10–40 microseconds depending on the laser parameters. This high supersaturation causes a very large nucleation density on the surface as compared to Molecular Beam Epitaxy or Sputtering Deposition. This nucleation density increases the smoothness of the deposited film.

In PLD, [depending on the deposition parameters above] three growth modes are possible:

- *Step-flow growth* – All substrates have a miscut associated with the crystal. These miscuts give rise to atomic steps on the surface. In step-flow growth, atoms land on the surface and diffuse to a step edge before they have a chance to nucleated a surface island. The growing surface is viewed as steps travelling across the surface. This growth mode is obtained by deposition on a high miscut substrate, or depositing at elevated temperatures
- *Layer-by-layer growth* – In this growth mode, islands nucleate on the surface until a critical island density is reached. As more material is added, the islands continue to grow until the islands begin to run into each other. This is known as coalescence. Once coalescence is reached, the surface has a large density of pits. When additional material is added to the surface the atoms diffuse into these pits to complete the layer. This process is repeated for each subsequent layer.
- *3D growth* – This mode is similar to the layer-by-layer growth, except that once an island is formed an additional island will nucleate on top of the 1st island. Therefore the growth does not persist in a layer by layer fashion, and the surface roughens each time material is added.

History

Pulsed laser deposition is only one of many thin film deposition techniques. Other methods include molecular beam epitaxy (MBE), chemical vapor deposition (CVD), sputter deposition (RF, Magnetron, and ion beam). The history of laser-assisted film growth started soon after the technical realisation of the first laser in 1960 by Maiman. Smith and Turner utilised a ruby laser to deposit the first thin films in 1965, three years after Breech and Cross studied the laser-vaporization and excitation of atoms from solid surfaces. However, the deposited films were still inferior to those obtained by other techniques such as chemical vapor deposition and molecular beam epitaxy. In the early 1980s, a few research groups (mainly in the former USSR) achieved remarkable results on manufacturing of thin film structures utilising laser technology. The breakthrough came in 1987 when Dijkkamp and Venkatesan were able to laser deposit a thin film of $YBa_2Cu_3O_7$, a high temperature superconductive material, which was of superior quality to that of films deposited with alternative techniques. Since then, the technique of Pulsed Laser Deposition has been utilised to fabricate high quality crystalline films. The deposition of ceramic oxides, nitride films, metallic multilayers and various superlattices has been demonstrated. In the 1990s the development of new laser technology, such as lasers with high repetition rate and short pulse durations, made PLD a very competitive tool for the growth of thin, well defined films with complex stoichiometry.

Technical Aspects

There are many different arrangements to build a deposition chamber for PLD. The target material which is evaporated by the laser is normally found as a rotating disc attached to a support. However, it can also be sintered into a cylindrical rod with rotational motion and a translational up and down movement along its axis. This special configuration allows not only the utilisation of a synchronized reactive gas pulse but also of a multicomponent target rod with which films of different multilayers can be created.

Some factors that influence the deposition rate:

- Target material
- Pulse energy of laser
- Distance from target to substrate
- Type of gas and pressure in chamber (oxygen, argon, etc.)

Low Pressure CVD

Epitaxy means "on top" or "assigned to", and represents a process in which a layer is created on top of another layer and inherits its crystal structure. If the deposited layer is of the same material as the substrate one speaks of homoepitaxy, if it's another material it's so-called heteroepitaxy. The most significant process in the homoepitaxy is the deposition of silicon on silicon, in heteroepitaxy usually a silicon layer is deposited on an insulator such as oxide (Silicon On Insulator: SOI). Homoepitaxy: Depending on the process, the wafers can be delivered from the wafer manufacturer with an epitaxial layer (e.g. for CMOS technology), or the chip manufacturer has to make it himself (for example in the bipolar technology).

As a gas for generating the epitactical layer, pure hydrogen is used in conjunction with silane (SiH_4), dichlorosilane (SiH_2Cl_2) or silicon tetrachloride ($SiCl_4$). At about 1000 °C, the gases cleave off the silicon, which is deposited on the wafer surface. The silicon inherits the structure of the substrate and is growing, for energy reasons, layer by layer successively on. To not grow up a polycrystalline silicon, one must always prevail a shortage of silicon atoms, e.g. it is always slightly less silicon available as material could actually grow up. When silicon tetrachloride is used, the reaction proceeds in two steps:

$$SiCl_4 + H_2 \rightarrow SiCl_2 + 2HCl2\ SiCl_2 \rightarrow Si + SiCl_4$$

In order to inherit the substrate's orientation the surface must be absolutely clear. So one can utilise the equilibrium reaction. Both reactions can occur in the other direction, depending on the ratio of the gases. If there is only few hydrogen in the atmosphere, as in the trichlorosilane process for the purification of raw silicon, material is removed from the silicon wafer surface due to the high chlorine concentration. Only with increasing concentration of hydrogen growth is achieved.

With $SiCl_4$ the deposition rate is approximately 1 to 2 microns per minute. Since the monocrystalline silicon grows only on the bare surface, certain areas can be masked with oxide where the silicon grows as polycrystalline silicon. This polysilicon, however, is etched very easily compared to single-crystalline silicon through the backward-running reaction. Diborane (B_2H_6) or phosphine (PH_3) are added to the process gases, to create doped layers, since the doping gases decompose at high temperatures and the dopants are incorporated in the crystal lattice.

The process to create home-epitactical layers is realised under vacuum atmosphere. Therefore the process chamber is heated to 1200 °C to remove the native oxide, which is always present on the silicon surface. As mentioned above, due to a low hydrogen concentration there occurs a back etch on the silicon surface. This can be used to clean the surface before the actual process starts. If the gas concentration is varied post this cleaning the deposition begins.

Illustration of a Barrel Reactor for Epitactical Processes

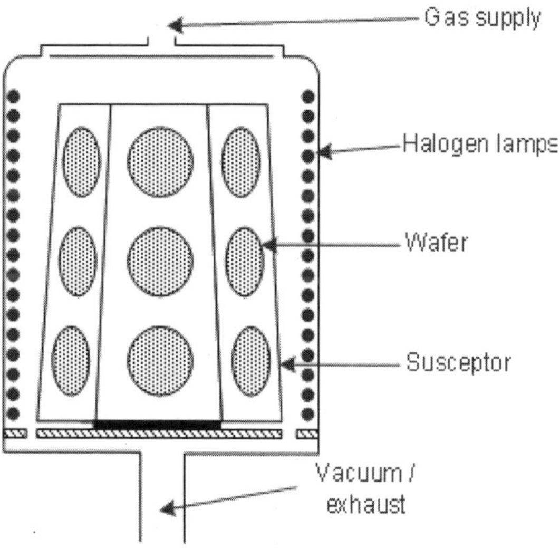

Barrel reactor

Due to the high process temperatures there's a diffusion of dopants in the substrate or impurities, which have been used in earlier processes, can move to the substrate. If SiH_2Cl_2 or SiH_4 are used there's no need for such high temperatures, so these gases are used primarily. To achieve the etch back process to clean the surface, HCl has to be added separately. The disadvantage of this silanes is that they form germs in the atmosphere right before deposition, and thus the quality of the layer is not as good as with $SiCl_4$.

CVD Process: Chemical Vapor Deposition

There is often a need of layers which can't be created right from the substrate. To deposit layers of silicon nitride or silicon oxynitride one has to use gases which contain all necessary components. The gases are decomposed via thermal energy. That's the principle of the chemical vapor phase deposition: CVD. The wafer surface doesn't react with the gases but serves as bottom layer. Depending on the process parameters - pressure, temperature - the CVD method can be devided in different methods whose layers differ in density and coverage. If the growth on horizontal surfaces is as high as on vertical surfaces the deposition is conform.

The conformity K is the ratio of vertical and horizontal growth, $K = R_v / R_h$. If the deposition is not ideal, the conformity is less than 1 (eg $R_v/R_h = 1/2 \rightarrow K = 0.5$). A high conformity can only be achieved by high process temperatures.

Imaginable Profiles

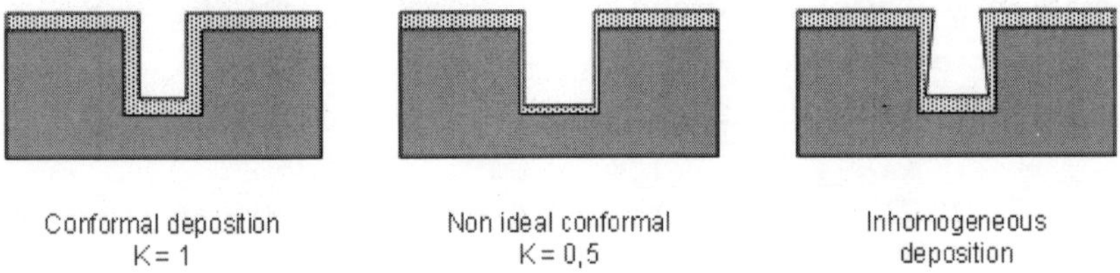

APCVD: Atmospheric Pressure CVD

APCVD is a CVD method at normal pressure (atmospheric pressure) which is used for deposition of doped and undoped oxides. The deposited oxide has a low density and the coverage is moderate due to a relatively low temperature. Because of improved tools, the APCVD undergoes a renaissance. The high wafer throughput is a big advantage of this process.

As process gases silane SiH_4 (highly deluted with nitrogen N_2) and oxygen O_2 are used. The gases are decomposed thermal at about 400 °C and react with each other to form the desired film.

$$SiH_4 + O_2 \rightarrow SiO_2 + 2H_2 \text{ (T = 430 °C, p = } 10^5 \text{ °Pa)}$$

Added ozone O_3 can cause a better conformity because it improves the movability of the accumulated particles. The oxide is porous and electrical instable and can be densified by a high temperature process.

To avoid edges which can result in difficulties at the deposition of additional layers, phosphorus silicate glass (PSG) is used for interlayers. Therefore phosphine is added to SiH_4 and O_2, so that the deposited oxide contains 4 to 8 % phosphorus. A high amount of phosphorus leads to a high increase of the flow properties, however, phosphoric acid can be formed which corrodes aluminum (conductor paths).

Because annealing affects earlier processes (e.g. doping) only short tempering is done with powerful argon lamps (several hundrets kW, less than 10s, T = 1100 °C) instead of annealing in longsome furnace processes.

Analog to PSG boron can be added simultaneously (boron phosphorus silicate glass, BPSG, 4 % B and 4 % P).

Illustration of a Horizontal APCVD Reactor

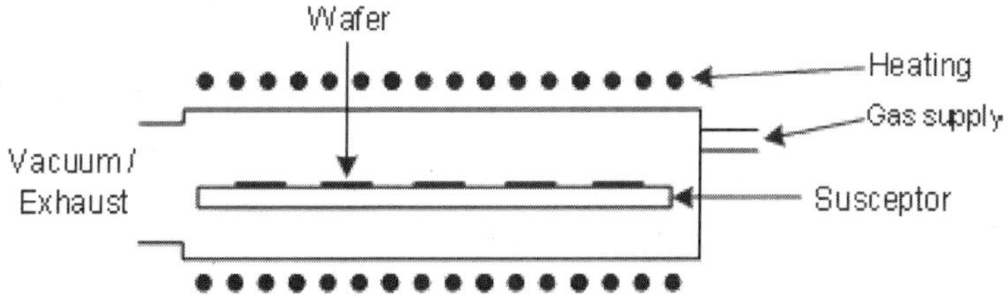

Horizontal reactor

LPCVD: Low Pressure CVD

In LPCVD a vacuum is used. Thin films of silicon nitride (Si_3N_4), silicon oxynitride (SiON), SiO_2 und tungsten (W) can be created. LPCVD processes enable a high conformity of almost 1. This is because of the low pressure of 10 to 100 Pa (atmospheric pressure = 100.000 Pa) which leads to a non-uniform movement of the particles. The particles dispread due to collisions and cover vertical surfaces as well as horizontal ones. The conformity is supported by a high temperature of up to 900 °C. Compared to APCVD the density and stability is very high.

The reactions for Si_3N_4, SiON, SiO_2 and tungsten are as follows:

a) Si_3N_4 (850 °C): $4NH_3 + 3SiH_2Cl_2 \rightarrow Si_3N_4 + 6HCl + 6H_2$

b) SiON (900 °C): $NH_3 + SiH_2Cl_2 + N_2O \rightarrow Si_3N_4$ + Nebenprodukte

c) SiO_2 (700 °C): $SiO_4C_8H_{20} \rightarrow SiO_2$ + Nebenprodukte

d) Wolfram (400 °C): $WF_6 + 3H_2 \rightarrow W + 6HF$

In contrast to gaseous precursors which are used for Si_3N_4, SiON and tungsten, liquid tetraethyl orthosilicate is used for SiO_2. Besides there are other liquid sources like DTBS ($SiH_2C_8H_{20}$) or tetramethylcyclotetrasiloxane (TMTCS, $Si_4O_4C_4H_{16}$).

A tungsten film can only be fabricated on bare silicon. Therefore silane has to be added if there is no silicon substrate.

Illustration of a LPCVD Reactor for TEOS Films

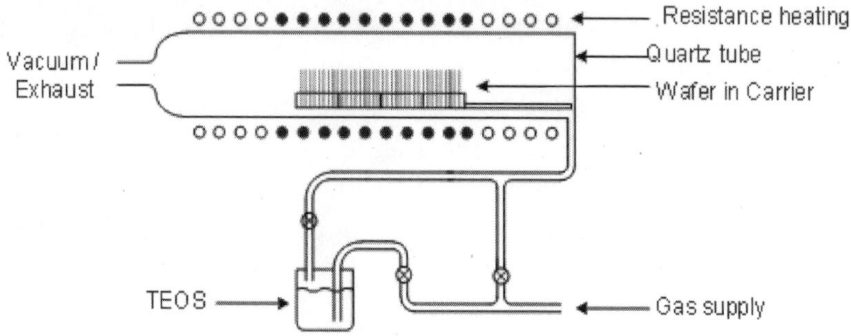

PECVD: Plasma Enhanced CVD

The PECVD takes place at 250 to 350 °C. Due to low temperatures the process gases can not be decomposited thermal. With a high frequency voltage, the gas is transformed into a plasma state. The plasma is energetic and disposes on the surface. Because metallization, such as aluminum, can not be exposed to high temperatures, the PECVD is used for SiO_2 and Si_3N_4 deposition on top of metal layers. Instead of SiH_2Cl_2 silane is used because it decomposes at lower temperature. The conformity is not as good as in LPCVD (0.6 to 0.8), however, the deposition rate is much higher (0.5 microns per minute).

Illustration of a PECVD Reactor

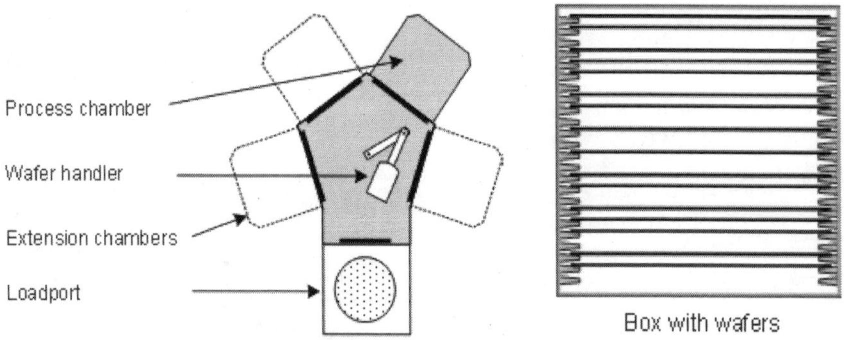

ALD: Atomic Layer Deposition

Atomic Layer Deposition (ALD) is a modified CVD process to manufacture thin films. The process uses several gases which are led into the process chamber alternating. Each gas reacts in such a way that the current surface is saturated, and therefore the reaction comes to a standstill. The alternative gas is able to react with this surface in the same way. Between the reactions of these gases the chamber is purged with an inert gas, like nitrogen or argon. A simple ALD process could look like this:

1. self-limiting reaction at the surface with first gas
2. purging with an inert gas
3. self-limiting reaction at the surface with second gas
4. purging with an inert gas

A specific example for an ALD process is the deposition of aluminum oxide, this can be realised with trimethylaluminum (TMA, C_3H_9Al) and water (H_2O).

First step is the elimination of hydrogen atoms which are bound to oxygen at the wafer surface. The methyl groups (CH_3) of TMA can react with the hydrogen to form methane (CH_4). The remaining molecules bond with the unsaturated oxygen.

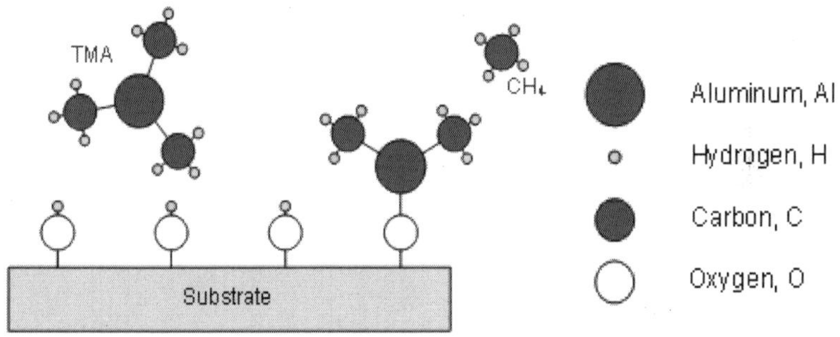

If these atoms are saturated, no more TMA molecules can react at the surface.

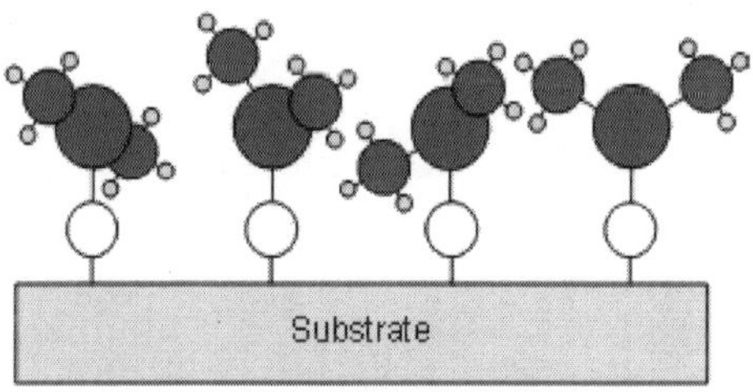

The chamber is purged and subsequent water steam is led into the chamber. Ever one hydrogen atom of the H_2O molecules now can react with the former deposited surface atoms to form methane, while the hydroxyl anion is bond to the aluminum atoms.

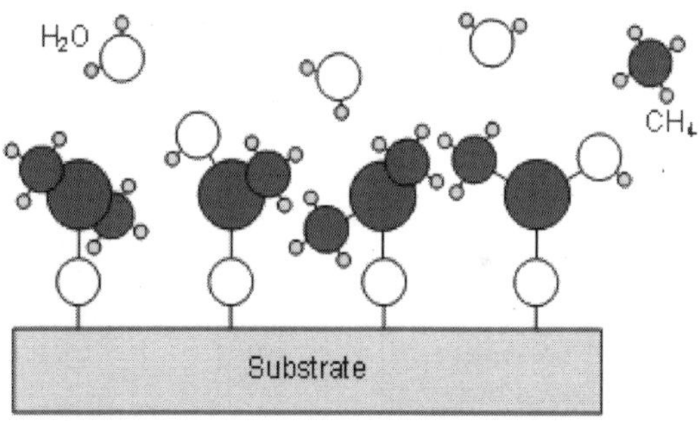

Hence, there are new hydrogen atoms at the surface which can react in a afterwards step with TMA like in the beginning.

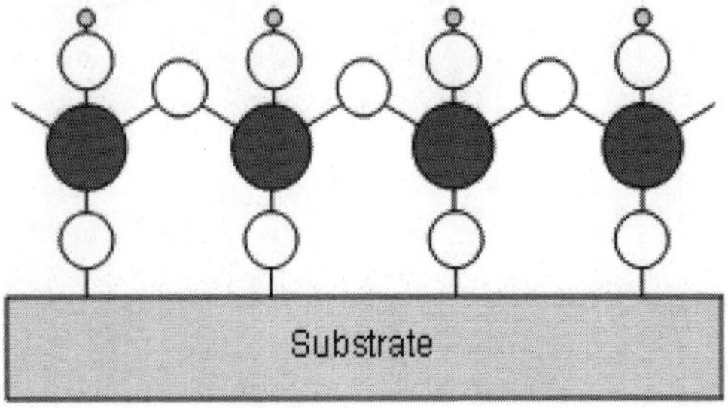

The atomic layer deposition provides significant advantages over other deposition techniques, and therefore it's a very important process to manufacture thin films. With ALD even 3-dimensional structures can be deposited very uniform. Insulating films are possible as well as conductive ones, which can be created on different substrates (semiconductors, polymers, ...). The film thickness can be controlled very precise by the number of cycles. Since the reactive gases are not led into the chamber simultaneously, they can not form germs right before the actual deposition. Thus the quality of the films is very high.

Atomic Layer CVD ALCVD

Atomic layer epitaxy (ALE) or Atomic Layer Chemical Vapor Deposition (ALCVD), now more generally called Atomic Layer Deposition (ALD), is a specialised form of epitaxy that typically deposit alternating monolayers of two elements onto a substrate. The crystal lattice structure achieved is thin, uniform, and aligned with the structure of the substrate. The reactants are brought to the substrate as alternating pulses with "dead" times in between. ALE makes use of the fact that the incoming material is bound strongly until all sites available for chemisorption are occupied. The dead times are used to flush the excess material. It is mostly used in semiconductor fabrication to grow thin films of thickness of the atomic order.

Technique

This technique was invented in 1970, then patented in 1977, by Dr. Tuomo Suntola, at the University of Helsinki in Finland. Dr. Suntola tried in fact to grow thin films of Zinc sulfide to fabricate electroluminescent flat panel displays. The main trick used for this technique is the use of a self-limiting chemical reaction to control in a very accurate way the thickness of the film deposited.

Compared to basic chemical vapour deposition for example, chemical reactants are pulsed alternatively in a reaction chamber and then chemisorb on to the surface of the substrate in order to form the monolayer. This is very clever because the reaction is very easy to set up and it doesn't require that many restrictions over the reactants, allowing the use of a wide range of materials.

ALD introduces two complementary precursors (e.g. $Al(CH_3)_3$ and H_2O) alternatively into the reaction chamber. Typically, one of the precursors will adsorb onto the substrate surface until it saturates the surface and further growth cannot occur until the second precursor is introduced. Thus the film thickness is controlled by the number of precursor cycles rather than the deposition time as is the case for conventional CVD processes. ALD allows for extremely precise control of film thickness and uniformity.

Nanoimprint Lithography

Nanoimprint lithography is a method of fabricating nanometre scale patterns. It is a simple nanolithography process with low cost, high throughput and high resolution. It creates patterns by mechanical deformation of imprint resist and subsequent processes. The imprint resist is typically a monomer or polymer formulation that is cured by heat or UV light during the imprinting. Adhesion between the resist and the template is controlled to allow proper release.

The term "Nanoimprint Lithography" (NIL) was coined in the scientific literature in 1996, when Prof. Stephen Chou and his students published a report in *Science*, although hot embossing (now taken as a synonym of NIL) of thermoplastics had been appearing in the patent literature for a few years already. Soon after the *Science* paper, many researchers developed different variations and implementations. At this point, nanoimprint lithography has been added to the International Technology Roadmap for Semiconductors (ITRS) for the 32 and 22 nm nodes.

Processes

There are many different types of nanoimprint lithography, but two of them are most important: thermoplastic nanoimprint lithography and photo nanoimprint lithography.

Thermoplastic Nanoimprint Lithography

Thermoplastic nanoimprint lithography (T-NIL) is the earliest nanoimprint lithography developed by Prof. Stephen Chou's group. In a standard T-NIL process, a thin layer of imprint resist (thermoplastic polymer) is spin coated onto the sample substrate. Then the mold, which has predefined topological patterns, is brought into contact with the sample and they are pressed together under certain pressure. When heated up above the glass transition temperature of the polymer, the pattern on the mold is pressed into the softened polymer film. After being cooled down, the mold is separated from the sample and the pattern resist is left on the substrate. A pattern transfer process (reactive ion etching, normally) can be used to transfer the pattern in the resist to the underneath substrate.

Alternatively, cold welding between two metal surfaces could also transfer low dimensional nanostructured metal without heating (especially for critical sizes less than ~10 nm),. Three-dimensional structures can be fabricated by repeating this procedure. The cold welding approach has the advantage of reducing surface contact contamination or defect due to no heating process, which is a main problem in the latest development and fabrication of organic electronic devices as well as novel solar cells.

Photo Nanoimprint Lithography

In photo nanoimprint lithography (P-NIL), a photo(UV) curable liquid resist is applied to the sample substrate and the mold is normally made of transparent material like fused silica. After the mold and the substrate are pressed together, the resist is cured in UV light and becomes solid. After mold separation, a similar pattern transfer process can be used to transfer the pattern in resist onto the underneath material. The use of a UV-transparent mold is difficult in a vacuum, because a vacuum chuck to hold the mold would not be possible.

Schemes

Full Wafer Nanoimprint: In a full wafer nanoimprint scheme, all the patterns are contained in a single nanoimprint field and will be transferred in a single imprint step. This allows a high throughput and uniformity. An at least 8-inch (203 mm) diameter full-wafer nanoimprint with high fidelity is possible.

To ensure the pressure and pattern uniformities of full wafer nanoimprint processes and prolong the mold lifetime, a pressing method utilising isotropic fluid pressure, named Air Cushion Press (ACP) by its inventors, is developed and being used by commercial nanoimprint systems.

Step and Repeat Nanoimprint

Nanoimprint can be performed in a way similar to the step and repeat optical lithography. The imprint field (die) is typically much smaller than the full wafer nanoimprint field. The die is repeatedly imprinted to the substrate with certain step size. This scheme is good for nanoimprint mold creation.

Applications

Nanoimprint lithography has been used to fabricate devices for electrical, optical, photonic and biological applications. For electronics devices, NIL has been used to fabricate MOSFET, O-TFT, single electron memory. For optics and photonics, intensive study has been conducted in fabrication of subwavelength resonant grating filter, polarizers, waveplate, anti-reflective structures, integrated photonics circuit and plasmonic devices by NIL. sub-10 nm nanofluidic channels had been fabricated using NIL and used in DNA strenching experiment. Currently, NIL is used to shrink the size of biomolecular sorting device an order of magnitude smaller and more efficient.

Benefits

A key benefit of nanoimprint lithography is its sheer simplicity. The single greatest cost associated with chip fabrication is the optical lithography tool used to print the circuit patterns. Optical lithography requires high powered excimer lasers and immense stacks of precision ground lens elements to achieve nanometre scale resolution. There is no need for complex optics or high-energy radiation sources with a nanoimprint tool. There is no need for finely tailored photoresists designed for both resolution and sensitivity at a given wavelength. The simplified requirements of the technology lead to its low cost.

Imprint lithography is inherently a three-dimensional patterning process. Imprint molds can be fabricated with multiple layers of topography stacked vertically. Resulting imprints replicate both layers with a single imprint step, which allows chip manufactures to reduce chip fabrication costs and improve product throughput. As mentioned above, the imprint material does not need to be finely tuned for high resolution and sensitivity. A broader range of materials with varying properties are available for use with imprint lithography. The increased material variability gives chemists the freedom to design new functional materials rather than sacrificial etch resistant polymers. A functional material may be imprinted directly to form a layer in a chip with no need for pattern transfer into underlying materials. The successful implementation of a functional imprint material would result in significant cost reductions and increased throughput by eliminating many difficult chip fabrication processing steps.

Concerns

The key concerns for nanoimprint lithography are overlay, defects, template patterning and template wear. However, recently Kumar et al. have shown that amorphous metals (metallic glasses) can be patterned on sub-100 nm scale, which can significantly reduce the template cost.

Overlay

The current overlay 3 sigma capability is 10 nm. Overlay has a better chance with step-and-scan approaches as opposed to full-wafer imprint.

Defects

As with immersion lithography, defect control is expected to improve as the technology matures. Defects from the template with size below the post-imprint process bias can be eliminated. Other defects would require effective template cleaning and/or the use of intermediate polymer stamps. When vacuum is not used during the imprint process, air can get trapped, resulting in bubble defects. This is because the imprint resist layer and the template or stamp features are not perfectly flat. There is an elevated risk when the intermediate or master stamp contains depressions (which are especially easy air traps), or when the imprint resist is dispensed as droplets just before imprinting, rather than pre-spun onto the substrate. Sufficient time must be allowed for the air to escape. Another issue is adhesion between stamp and resist. High adhesion (sticking) may delaminate resist, which than stays on stamp. This effect degrades pattern, reduces yield and damages stamp. It can be mitigated by employing an FDTS antistiction layer on a stamp.

Template Patterning

High resolution template patterning can currently be performed by electron beam lithography or focused ion beam patterning; however at the smallest resolution, the throughput is very slow. As a result, optical patterning tools will be more helpful if they have sufficient resolution. Such an approach has been successfully demonstrated by Greener et al. whereby robust templates were rapidly fabricated by optical patterning of a photoresist-coated metal substrate through a photomask. Other patterning techniques (including even double patterning) may also be used. Kumar and Schroers at Yale developed the nanopatterning of amorphous metals which can be used as inexpensive templates for nanoimprinting.

Template Wear

The use of substantial pressure to not only contact but also penetrate a layer during imprinting accelerates the wear of imprint templates compared to other types of lithographic masks. Template wear is reduced with proper use of an anti-adhesion FDTS monolayer coating on a stamp.

Other

Future applications of nanoimprint lithography may involve the use of porous low-k materials. These materials are not stiff and, as part of the substrate, are readily damaged mechanically by the pressure of the imprint process.

Removal of Residual Layers

A key characteristic of nanoimprint lithography (except for electrochemical nanoimprinting) is the residual layer following the imprint process. It is preferable to have thick enough residual layers to support alignment and throughput and low defects. However, this renders the nanoimprint lithography step less critical for critical dimension (CD) control than the etch step used to remove the residual layer. Hence, it is important to consider the residual layer removal an integrated part of the overall nanoimprint patterning process.

In a sense, the residual layer etch is similar to the develop process in conventional lithography. It has been proposed to combine photolithography and nanoimprint lithography techniques in one step in order to eliminate the residual layer.

Proximity Effects

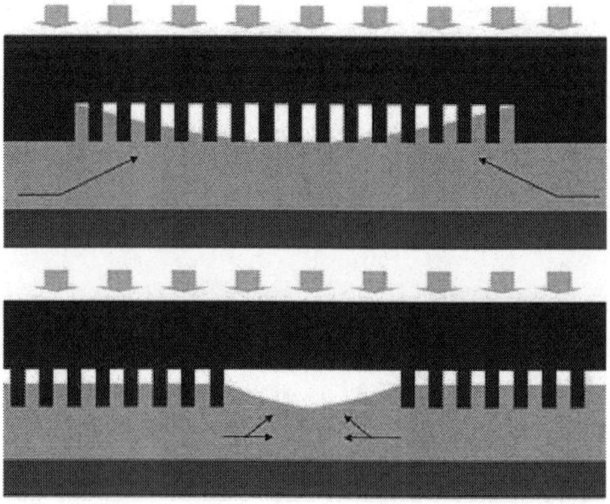

Figure: *Nanoimprint proximity effect.* Top: *Array of depressions is more quickly filled at the edge than the centre, resulting in less imprinting at the centre of the array.* Bottom: *The wide space between two groups of protrusions tends to be filled slower than the narrow spaces between the protrusions, resulting in the formation of holes in the unpatterned area.*

Nanoimprint lithography relies on displacing polymer. This could lead to systematic effects over long distances. For example, a large, dense array of protrusions will displace significantly more polymer than an isolated protrusion. Depending on the distance of this isolated protrusion from the array, the isolated feature may not imprint correctly due to polymer displacement and thickening. Resist holes can form in between groups of protrusions. Likewise, wider depressions in the template do not fill up with as much polymer as narrower depressions, resulting in misshapen wide lines. In addition, a depression at the edge of a large array fills up much earlier than one located in the centre of the array, resulting in within-array uniformity issues.

3D-Patterning

A unique benefit of nanoimprint lithography is the ability to pattern 3D structures, such as damascene interconnects and T-gates, in fewer steps than required for conventional lithography. This is achieved by building the T-shape into the protrusion on the template.

Alternative Approaches

Electrochemical Nanoimprinting

Electrochemical nanoimprinting can be achieved using a stamp made from a superionic conductor such as silver sulfide. When the stamp is contacted with metal, electrochemical etching can be carried out with an applied voltage. The electrochemical reaction generates metal ions which move from the original film into the stamp. Eventually all the metal is removed and the complementary stamp pattern is transferred to the remaining metal.

Laser Assisted Direct Imprint

Laser assisted direct imprint (LADI) is a rapid technique for patterning nanostructures in solid substrates and it does not require etching. A single or multiple excimer laser pulses melt a thin surface layer of substrate material, and a mold is embossed into the resulting liquid layer. A variety of structures with resolution better than 10 nm have been imprinted into silicon using LADI, and the embossing time is less than 250 ns. The high resolution and speed of LADI, attributed to molten silicon's low viscosity (one-third that of water), could open up a variety of applications and be extended to other materials and processing techniques.

The Future of Nanoimprint

Nanoimprint lithography is a simple pattern transfer process that is neither limited by diffraction nor scattering effects nor secondary electrons, and does not require any sophisticated radiation chemistry. It is also a potentially simple and inexpensive technique. However, a lingering barrier to nanometre-scale patterning is the current reliance on other lithography techniques to generate the template. It is possible that self-assembled structures will provide the ultimate solution for templates of periodic patterns at scales of 10 nm and less. It is also possible to resolve the template generation issue by using a programmable template in a scheme based on double patterning.

As of October 2007, Toshiba is the only company to have validated nanoimprint lithography for 22 nm and beyond. What is more significant is that nanoimprint lithography is the first sub-30 nm lithography to be validated by an industrial user.

5

XRay Lithography XRL and LIGA

X-ray lithography, is a process used in electronic industry to selectively remove parts of a thin film. It uses X-rays to transfer a geometric pattern from a mask to a light-sensitive chemical photoresist, or simply “resist,” on the substrate. A series of chemical treatments then engraves the produced pattern into the material underneath the photoresist.

Transmission from a clear mask feature

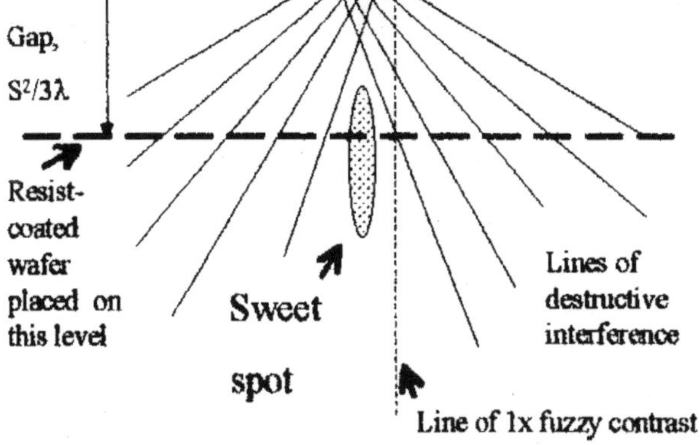

Mechanisms

X-ray lithography originated as a candidate for next-generation lithography for the semiconductor industry, with batches of microprocessors successfully produced. Having short wavelengths (below 1 nm), X-rays overcome the diffraction limits of optical lithography, allowing smaller feature sizes. If the X-ray source isn't collimated, as with a synchrotron radiation, elementary collimating mirrors or diffractive lenses are used in the place of the refractive lenses used in optics. The X-rays illuminate a mask placed in proximity of a resist-coated wafer. The X-rays are broadband, typically from a compact synchrotron radiation source, allowing rapid exposure. Deep X-ray lithography (DXRL) uses yet shorter wavelengths on the order of 0.1 nm and modified procedures such as the LIGA process, to fabricate deep and even three-dimensional structures.

The mask consists of an X-ray absorber, typically of gold or compounds of tantalum or tungsten, on a membrane that is transparent to X-rays, typically of silicon carbide or diamond. The pattern on the mask is written by direct-write electron beam lithography onto a resist that is developed by conventional semiconductor processes. The membrane can be stretched for overlay accuracy.

Most X-ray lithography demonstrations have been performed by copying with image fidelity (without magnification) on the line of fuzzy contrast as illustrated in the figure. However, with the increasing need for high resolution, X-ray lithography is now performed on what is called the "sweet spot", using local "demagnification by bias". Dense structures are developed by multiple exposures with translation. The advantages of using 3x demagnification include, the mask is more easily fabricated, the mask to wafer gap is increased, and the contrast is higher. The technique is extensible to dense 15 nm prints.

X-rays generate secondary electrons as in the cases of extreme ultraviolet lithography and electron beam lithography. While the fine pattern definition is due principally to secondaries from Auger electrons with a short path length, the primary electrons will sensitize the resist over a larger region than the X-ray exposure. While this does not affect the pattern pitch resolution, which is determined by wavelength and gap, the image exposure contrast (max-min)/(max+min) is reduced because the pitch is on the order of the primary photo-electron range. The sidewall roughness and slopes are influenced by these secondary electrons as they can travel few micrometres in the area under the absorber, depending on exposure X-ray energy. Several prints at about 30 nm have been published.

Another manifestation of the photoelectron effect is exposure to X-ray generated electrons from thick gold films used for making daughter masks. Simulations suggest that photoelectron generation from the gold substrate may affect dissolution rates.

Photoelectrons, Secondary Electrons and Auger Electrons

Secondary electrons have energies of 25 eV or less, and can be generated by any ionizing radiation (VUV, EUV, X-rays, ions and other electrons). Auger electrons have energies of hundreds of electronvolts. The secondaries (generated by and outnumbering the Auger and primary photoelectrons) are the main agents for resist exposure.

The relative ranges of photoelectron primaries and Auger electrons depend on their respective energies. These energies depend on the energy of incident radiation and on the composition of the resist. There is considerable room for optimum selection (reference 3 of the article). When Auger electrons have lower energies than primary photoelectrons, they have shorter ranges. Both decay to secondaries which interact with chemical bonds. When secondary energies are too low, they fail to break the chemical bonds and cease to affect print resolution. Experiments prove that the combined range is less than 20 nm. On the other hand, the secondaries follow a different trend below H"30 eV: the lower the energy, the longer the mean free path though they are not then able to affect resist development.

As they decay, primary photo-electrons and Auger electrons eventually become physically indistinguishable (as in Fermi–Dirac statistics) from secondary electrons. The range of low-energy secondary electrons is sometimes larger than the range of primary photo-electrons or of Auger electrons. What matters for X-ray lithography is the effective range of electrons that have sufficient energy to make or break chemical bonds in negative or positive resists.

Lithographic Electron Range

X-rays do not charge. The relatively large mean free path (~20 nm) of secondary electrons hinders resolution control at nanometre scale. In particular, electron beam lithography suffers negative charging by incident electrons and consequent beam spread which limits resolution. It is difficult therefore to isolate the effective range of secondaries which may be less than 1 nm.

The combined electron mean free path results in an image blur, which is usually modelled as a Gaussian function (where σ = blur) that is convolved with the expected image. As the desired resolution approaches the blur, the dose image becomes broader than the aerial image of the incident X-rays. The blur that matters is the latent image that describes the making or breaking of bonds during the exposure of resist. The developed image is the final relief image produced by the selected high contrast development process on the latent image.

The range of primary, Auger, secondary and ultralow energy higher-order generation electrons which print (as STM studies proved) can be large (tens of nm) or small (nm), according to various cited publications. Because this range is not a fixed number, it is hard to quantify. Line edge roughness is aggravated by the associated uncertainty. Line edge roughness is supposedly statistical in origin and only indirectly dependent on mean range. Under commonly practiced lithography conditions, the various electron ranges can be controlled and utilised.

Charging

X-rays carry no charge, but at the energies involved, Auger decay of ionized species in a specimen is more probable than radiative decay. High-energy radiation exceeding the ionization potential also generates free electrons which are negligible compared to those produced by electron beams which are charged. Charging of the sample following ionization is an extremely weak possibility when it cannot be guaranteed the ionized electrons leaving the surface or remaining in the sample are adequately balanced from other sources in time. The energy transfer to electrons as a result of ionizing radiation results in separated positive and negative charges which quickly recombine due partly to the long range of the Coulomb force. Insulating films like gate oxides and resists have been observed to charge to a positive or negative potential under electron-beam irradiation. Insulating films are eventually neutralized locally by space charge (electrons entering and exiting the surface) at the resist-vacuum interface and Fowler-Nordheim injection from the substrate. The range of the electrons in the film can be affected by the local electric field. The situation is complicated by the presence of holes (positively charged electron vacancies) which are generated along with the secondary electrons, and which may be expected to follow them around. As neutralization proceeds, any initial charge concentration effectively starts to spread out. The final chemical state of the film is reached after neutralization is completed, after all the electrons have eventually slowed down. Usually, excepting X-ray steppers, charging can be further controlled by flood gun or resist thickness or charge dissipation layer.

Extreme Ultraviolet Lithography

Extreme ultraviolet lithography (also known as *EUV* or *EUVL*) is a next-generation lithography technology using an extreme ultraviolet (EUV) wavelength, currently expected to be 13.5 nm.

EUVL Light Source

Neutral atoms or condensed matter cannot emit EUV radiation. For matter to emit it, ionization must take place first. EUV light can only be emitted by electrons which are bound to multicharged positive ions; for example, to remove an electron from a +3 charged carbon ion (three electrons already removed) requires about 65 eV. Such electrons are more tightly bound than typical valence electrons. The thermal production of multicharged positive ions is only possible in a hot dense plasma, which itself strongly absorbs EUV. The Xe or Sn plasma sources for EUV lithography are either discharge-produced or laser-produced. Discharge-produced plasma is made by lightning bolt's worth of electric current through a tin vapor. Laser-produced plasma is made by microscopic droplets of molten tin heated by powerful laser. Laser-produced plasma sources (e.g., ASML's NXE:3300B stepper) outperform discharge-produced plasma sources. Power output exceeding 100 W is a requirement for sufficient throughput. While state-of-the-art 193 nm ArF excimer lasers offer intensities of 200 W/cm^2, lasers for producing EUV-generating plasmas need to be much more intense, on the order of 10^{11} W/cm^2. This indicates the enormous energy burden imposed by switching from generating 193 nm light (laser output approaching 100 W) to generating EUV light (required laser or equivalent power source output exceeding 10 kW). An EUV source driven by a 200 kW CO_2 laser with ~10% wall plug efficiency consumes an electrical power of ~2 MW, while a 100 W ArF immersion laser with ~1% wall plug efficiency consumes an electrical power of ~10 kW.

A further characteristic of the plasma-based EUV sources under development is that they are not even partially coherent, unlike the KrF and ArF excimer lasers used for current optical lithography. Further power reduction (energy loss) is expected in converting incoherent sources (emitting in all possible directions at many independent wavelengths) to partially coherent (emitting in a limited range of directions within a narrow band of wavelengths) sources by filtering (unwanted wavelengths and directions). On the other hand, coherent light poses a risk of monochromatic reflection interference and mismatch of multilayer reflectance bandwidth.

As of 2008, the development tools had a throughput of 4 wafers per hour with a 120 W source. For a 100 WPH requirement, therefore, a 3 kW source would be needed, which is not available in the foreseeable future. However, EUV photon count is determined by the number of electrons generated per photon which are collected by a photodiode; since this is essentially the highly variable secondary yield of the initial photoelectron, the dose measurement will be impacted by high variability. In fact, data by Gullikson et al. indicated ~10% natural variation of the photocurrent responsivity. More recent data for silicon photodiodes remain consistent with this assessment. Calibration of the EUV dosimetre is a nontrivial unsolved issue. The secondary electron number variability is the well-known root cause of noise in avalanche photodiodes.

The highly relativistic vacuum tubes free-electron lasers and synchrotron radiation sources can give better light quality than material sources can, though high intensity may require development work. Several dedicated industrial synchrotron light facilities have been built, and their applications include semiconductor device fabrication. Free electron lasers offer light that is monochromatic and coherent, as well as narrow in space and angle spread. Both also offer a continuous range of available wavelengths, allowing seamless progress into the X-ray band.

At SPIE 2014, TSMC reported that the 200 kW CO_2 laser for their NXE:3100 EUV tool light source had a misalignment problem. The laser was supposed to focus on a tin droplet which absorbs the power to generate EUV light. Missing the droplet directed the power elsewhere, leading to component damage, and some downtime.

EUVL Optics

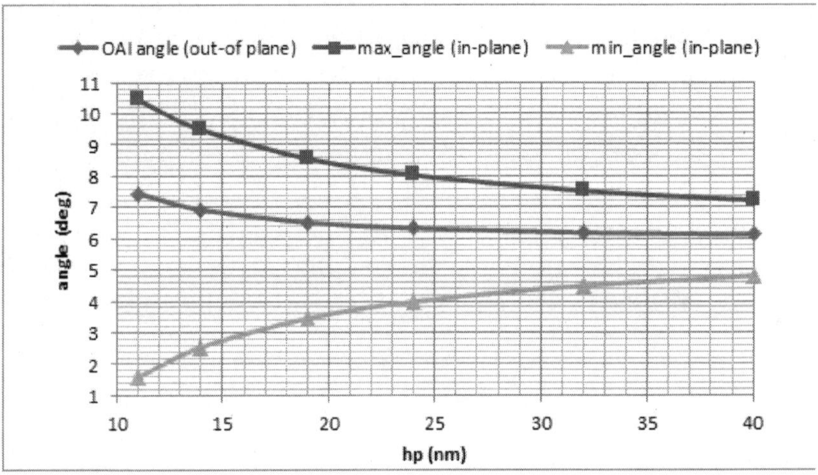

Figure: *EUV angle of incidence for different pitches. Due to the off-axis illumination angle, a 4x projection system exhibits two different incidence angles on the multilayer mask for dipole illuminations optimized for different pattern pitches; the tighter pitches exhibit larger angle differences. The central ray angle of incidence here is 6°.*

EUVL is a significant departure from the deep ultraviolet lithography used today. All matter absorbs EUV radiation. Hence, EUV lithography needs to take place in a vacuum. All the optical elements, including the photomask, must make use of defect-free Mo/Si multilayers which act to reflect light by means of interlayer interference; any one of these mirrors will absorb around 30% of the incident light. This limitation can be avoided in maskless interference lithography systems. However, the latter tools are restricted to producing periodic patterns only.

The pre-production EUVL systems built to date contain at least two condenser multilayer mirrors, six projection multilayer mirrors, and a multilayer object (mask). Since the optics already absorbs 96% of the available EUV light, the ideal EUV source will need to be sufficiently bright. EUV source development has focused on plasmas generated by laser or discharge pulses. The mirror responsible for collecting the light is directly exposed to the plasma and is therefore vulnerable to damage from the high-energy ions and other debris. This damage associated with the high-energy process of generating EUV radiation has precluded the successful implementation of practical EUV light sources for lithography.

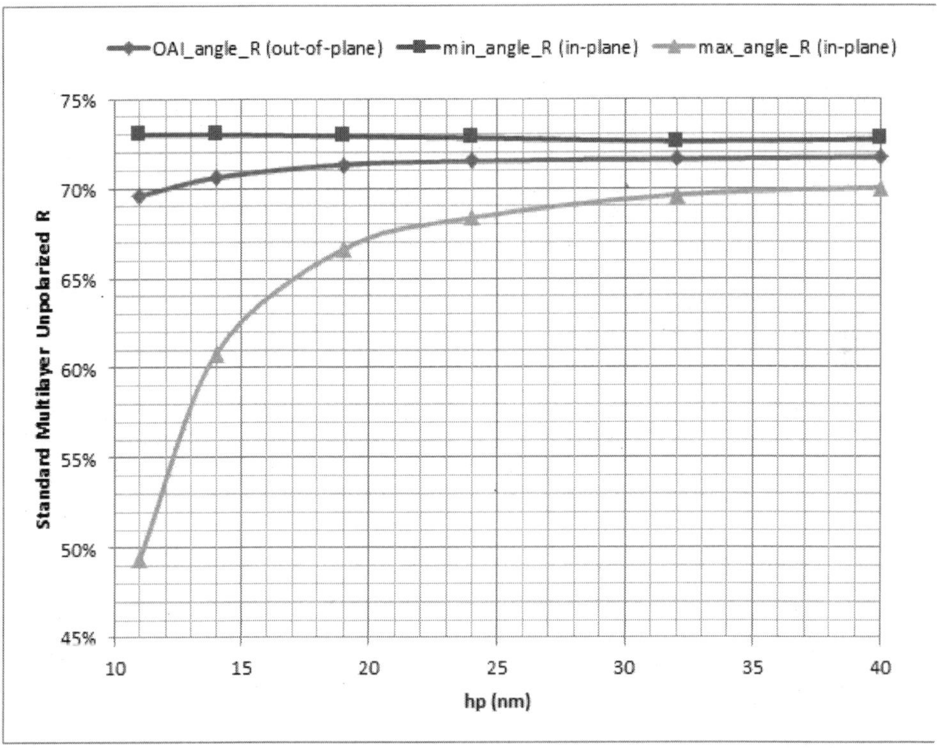

Figure: *EUV multilayer OAI reflectivity vs. half-pitch. Due to the off-axis illumination angle, a 4x projection system exhibits two different incidence angles on the multilayer mask for dipole illuminations optimized for different pattern pitches. As a result, the tighter pitches exhibit larger reflectivity differences. The central ray angle of incidence here is 6°.*

The wafer throughput of an EUVL exposure tool is a critical metric for manufacturing capacity. Given that EUV is a technology requiring high vacuum, the throughput is limited (aside from the source power) by the transfer of wafers into and out of the tool chamber, to a few wafers per hour. Another aspect of the pre-production EUVL tools is the off-axis illumination (at an angle of 6 degrees) on a multilayer mask. The resulting asymmetry in the diffraction pattern causes shadowing effects which degrade the pattern fidelity. EUVL's shorter wavelength also increases flare, resulting in less than perfect image quality and increased line width roughness. Heating per feature volume (e.g., 20 nm cube) is higher per EUV photon compared to a DUV photon, due to higher absorption in resist. In addition, EUV lithography results in more heating due to the vacuum environment, in contrast to the water cooling environment of immersion lithography.

Heating is also a particularly serious issue for multilayer mirrors used, because, as EUV is absorbed within a thin distance from the surface, the heating density is higher. As a result, water cooling is expected to be used for the high heating load; however, the resulting vibration is a concern.

Heating of the EUV mask pellicle (film temperature up to 750 K for 80 W incident power) is a significant concern, due to the resulting deformation and decrease of transmission.

A recent study by NIST and Rutgers University found that multilayer optics contamination was highly affected by the resonant structure of the EUV mirror influencing the photoelectron generation and secondary electron yield.

Since EUV is highly absorbed by all materials, even EUV optical components inside the lithography tool are susceptible to damage, mainly manifest as observable ablation. Such damage is a new concern specific to EUV lithography, as conventional optical lithography systems use mainly transmissive components and electron beam lithography systems do not put any component in the way of electrons, although these electrons end up depositing energy in the exposed sample substrate.

In 2012, the Laser-Laboratorium Göttingen and KLA-Tencor reported that a Ru-capped Mo/Si multilayer could be damaged by a single pulse (16° incidence at 13.5 nm) as low as 30 mJ/cm^2, and the damage threshold can be lowered ~60% with ten pulses. This was attributed to the cumulative probability of defect occurrence with multiple pulses.

Another collaboration study in 2010, using Mo/Si multilayers with 42-44% reflectivity at ~28° incidence, started showing damage for single pulse at a level of ~45 mJ/cm^2.

EUV-Specific Overlay Issues

Because EUV operates in a vacuum and requires reflective optics, EUV lithography tools have special overlay concerns, recently studied by IMEC, along with ASML. Electrostatic chucks must be used instead of conventional vacuum chucks. Therefore the wafer clamping variability on the electrostatic chuck needs to be dealt with. A backside coating of 200 nm silicon nitride (which must be removed later to allow backside cooling and heatsinking) was found to be helpful. Other than this additional step, which also requires first protecting the device layers already patterned, zone alignment (using all alignment marks across the wafer, not a standard subset) also provided some improvement. The vacuum environment required by EUV also leads to heating of the wafer without much dissipation.

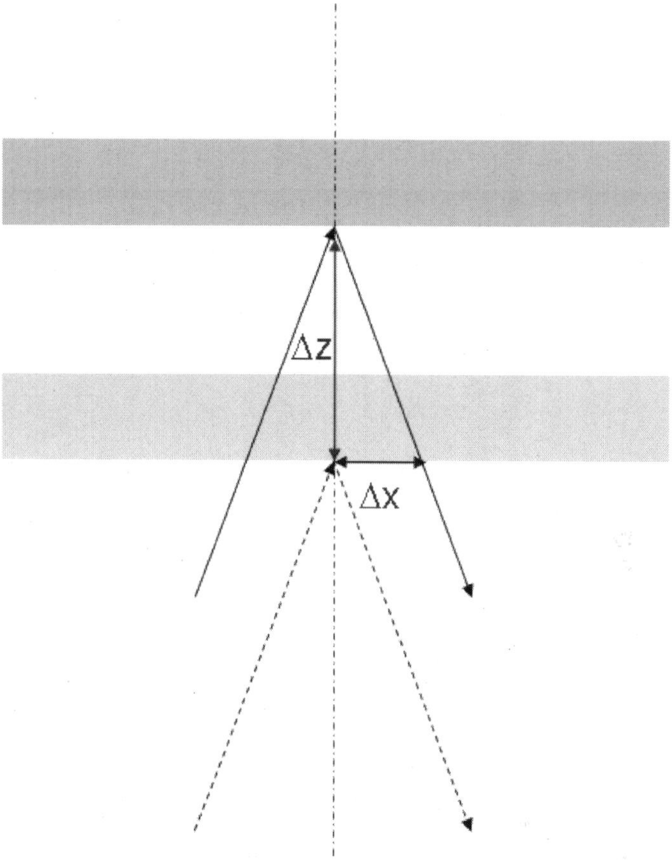

Figure: *EUV non-telecentricity. The off-axis reflection of light rays from different vertical positions of the reticle leads to different lateral displacements of the imaged feature on the wafer.*

A sacrificial first wafer was found to be necessary to stabilize the chuck temperature. Moreover, the local overlay corrections due to exposure heating requires the use of a second wafer. Thus, an extra wafer per lot is required for overlay stabilization in EUV lithography. The use of reflection causes wafer exposure position to be extremely sensitive to the reticle flatness and the reticle clamp. Reticle clamp cleanliness is therefore required to be maintained.

EUV Exposure of Photoresist

When an EUV photon is absorbed, photoelectrons and secondary electrons are generated by ionization, much like what happens when X-rays or electron beams are absorbed by matter. It has been estimated that about 4 secondary electrons on average are generated for every EUV photon, although the generation volume is not definite. These secondary electrons have energies of a few to tens of eV and travel tens of nanometres inside photoresist before initiating the desired chemical reaction. This is very similar to the photoelectron migration for the latent image formation in silver halide photographic films. A contributing factor for this rather large distance is the fact that polymers have significant amounts of free volume. In a recent actual EUV print test, it was found 30 nm spaces could not be resolved, even though the optical resolution and the photoresist composition were not the limiting factor.

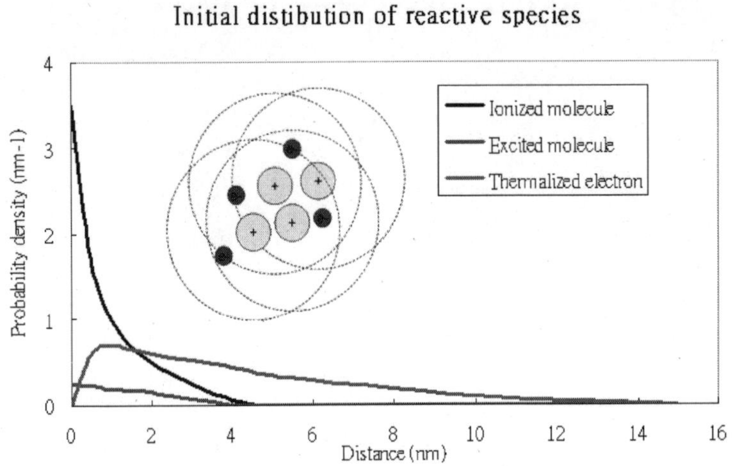

Figure: *Initial distribution of reactive species after EUV absorption. Molecules are excited and ionized within a few nanometres from the absorption point, and electrons are thermalized within 20 nanometres from the absorption point. The inset picture shows the multispur effect, where several electron-ion pairs generated by the EUV photon may interact with one another.*

In particular, for photoresists utilising chemical amplification for higher throughput:

e^{-} + acid generator -> anion -> dissociated anion products

This reaction, also known as "electron attachment" or "dissociative electron attachment" is most likely to occur after the electron has essentially slowed to a halt, since it is easiest to capture at that point. The cross-section for electron attachment is inversely proportional to electron energy at high energies, but approaches a maximum limiting value at zero energy. On the other hand, it is already known that the mean free path at the lowest energies (few to several eV or less, where dissociative attachment is significant) is well over 10 nm, thus limiting the ability to consistently achieve resolution at this scale. In addition, electrons with energies < 20 eV are capable of desorbing hydrogen and fluorine anions from the resist, leading to potential damage to the EUV optical system.

EUV photoresist images often require resist thicknesses roughly equal to the pitch. This is not only due to EUV absorption causing less light to reach the bottom of the resist but also to forward scattering from the secondary electrons (similar to low-energy electron beam lithography). Conversely, thinner resist transmits a larger fraction of incident light allowing damage to underlying films, yet requires more dosage to achieve the same level of absorption. Since the photon absorption depth exceeds the electron escape depth, as the released electrons eventually slow down, they dissipate their energy ultimately as heat.

An EUV dose of 1 mJ/cm^2 generates an equivalent photoelectron dose of 10.9 μC/cm^2. Current demonstration doses exceed 10 mJ/cm^2, or equivalently, 109 μC/cm^2 photoelectron dose.

The use of higher doses and/or reduced resist thicknesses to produce smaller features only results in increased irradiation of the layer underneath the photoresist. This adds another significant source of photoelectrons and secondary electrons which effectively reduce the image contrast. In addition, there is increased possibility of ionizing radiation damage to the layers below.

The extent of secondary electron and photoelectrons in blurring the resolution is dependent on factors such as dose, surface contamination, temperature, etc.

EUVL Defects

EUVL faces specific defect issues analogous to those being encountered by immersion lithography. Whereas the immersion-specific defects are due to unoptimized contact between the water and the photoresist, EUV-related defects are attributed to the inherently ionizing energy of EUV radiation. The first issue is positive charging, due to ejection of photoelectrons freed from the top resist surface by the EUV radiation. This could lead to electrostatic discharge or particle contamination as well as the device damage mentioned above. A second issue is contamination deposition on the resist from ambient or outgassed hydrocarbons, which results from EUV- or electron-driven reactions. A third issue is etching of the resist by oxygen, argon or other ambient gases, which have been dissociated by the EUV radiation or the electrons generated by EUV. Ambient gases in the lithography chamber may be used for purging and contamination reduction. These gases are ionized by EUV radiation, leading to plasma generation in the vicinity of exposed surfaces, resulting in damage to the multilayer optics and inadvertent exposure of the sample.

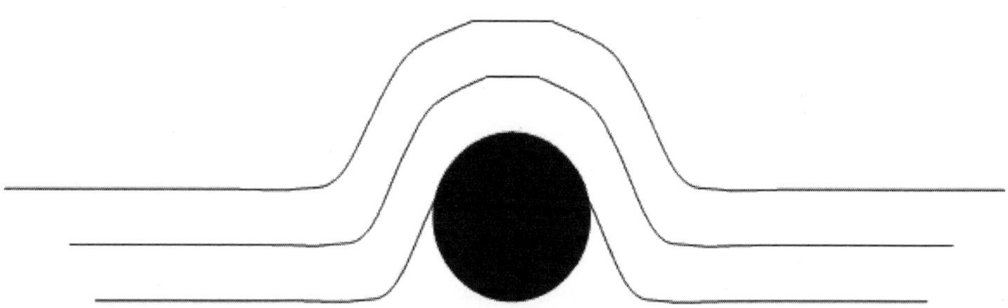

Figure: *Buried defect on EUV mask blank. A particle on the substrate distorts the multilayer deposited over it, the distortion becoming wider as more layers are added.*

Of course mask defects are also a known source of defects for EUVL. Reducing defects on extreme ultraviolet (EUV) masks is currently one of the most critical issues to be addressed for commercialization of EUV lithography. The defect core, namely the pit or particle, can originate either on the substrate, during multilayer deposition or on top of the multilayer stack. The printability of the final defect will depend on the phase change and the amplitude change of light at a given position. The net phase change and/or amplitude change adds to the intrinsic effect of the core defect and its influence on the growth of the multilayer stack during deposition. The buried defects are particularly insidious, and even 10 nm defects may be considered risky. The phase shift caused by an undetected 3 nm mask substrate flatness variation is sufficient to produce a printable defect. The principle behind this is a quarter-wavelength deviation from the flat surface produces a half-wavelength optical path difference after reflection. The light that is reflected from the flat surface is 180 degrees out of phase with the light reflected from the quarter-wavelength deviation. It has been shown that even a 1 nm deviation from flatness would lead to a substantial reduction (~20%) of the image intensity. In fact, defects of atomic scale height (0.3-0.5 nm) with 100 nm FWHM can still be printable by exhibiting 10% CD impact. Like a lens, any defect which effectively produces a phase shift scatters light outside the defect region. The amount of light that is scattered can be calculated. Furthermore, the edge of a phase defect will further reduce reflectivity by more than 10% if its deviation from flatness exceeds 3 degrees, due to the deviation from the target angle of incidence of 84 degrees with respect to the surface. Even if the defect height is shallow, the edge still deforms the overlying multilayer, producing an extended region where the multilayer is sloped. The more abrupt the deformation, the narrower the defect edge extension, the greater the loss in reflectivity.

Unexpected Resolution Limits

Given that EUV is a significant reduction in wavelength compared to current lithography wavelengths, one would expect significantly better resolution. However, the resolution is ultimately determined by other factors besides diffraction, such as the interaction volume in the image recording medium, i.e., the photoresist. As noted above, the low energy electrons released by EUV could blur the original EUV image. In addition, there are statistical effects, especially for feature areas less than 1500 square nanometres. The multilayer supporting EUV optics is also of limited bandwidth, posing scaling limits which would require further modification.

Resist Polymer Aggregates

Resists as polymers are well-known to have aggregates with sizes up to 80 nm. Even the high-resolution resist HSQ has aggregate size reduced down to only 15–20 nm. While the roughness of lines larger than the aggregate size are mildly affected by the aggregate size, below the aggregate size obviously the linewidth can be severely affected.

Resist Line Roughness

Minimum EUV dose to avoid resist line roughness For PMMA and novolac-based resists (not chemically amplified), the minimum EUV doses to avoid sidewall or base defects are shown here.

A model for resist line roughness caused by defects proposed in 1994 predicted that if minimum EUV dose levels for a given linewidth were not met, the resulting roughness from sidewall or base defects would be prohibitive. For 20 nm linewidth and below, the minimum dose easily exceeds 100 mJ/cm^2 for resists that are not chemically amplified.

Shot Noise

A dose sensitivity of 5 mJ/cm^2 implies only several thousand EUV photons or so accumulate in such a small area. With the natural Poisson distribution due to the random arrival times of the photons, there is an expected natural dose variation of at least a few percent 3 sigma, making the exposure process fundamentally uncontrollable for features less than about 40 nm. Increasing the dose will reduce the shot noise, but will also increase the flare dose and generate more free electrons. The free electrons will spread out before slowing to a stop. Since the free electron density is lower than the initial photon density, the shot noise is always effectively larger than expected from just considering the EUV dose.

In 2008, Intel calculated that for printing one billion 30 nm contacts, ± 16% dose error (@10 mJ/cm^2) is expected from the EUV shot noise. With acid counting, the fluctuation increases to ± 20%. This issue will affect 22 nm patterning integration. When one considers that within a 1 nm pixel, the shot noise is even more significant (>100% on 10 nm scale @10 mJ/cm^2), the origin of the line edge roughness (LER) issue in EUV lithography becomes clearer.

The 2D patterns often encountered in DRAM and logic microprocessors (including multiple pattern line cutting for 11 nm node complementary lithography) as well as the floating gate flash memory patterns with 2D isolation for charge trapping are more susceptible to shot noise than line-type features. It is because the 2D pattern (ideally rectangular) is defined by the number of photons in a limited area exposed above or below a certain threshold dose.

feature diameter (nm)	*minimum dose to avoid 5% dose error among 1 million features (mJ/cm^2)*	*targeted dose (mJ/cm^2)*	*throughput at targeted dose (300 mm WPH)*
40	11	5	-
28	22	10	6-60
20	42	15	50-125
14	86	20	125
10	169	20	165

A 5% dose error has been found to result in ~1 nm CD error. Although the minimum dose to avoid 5% dose error within a population of a million contacts is doubling every generation, the industry targeted dose is not keeping up. To at least make up the minimum dose, the throughput will be reduced by the same ratio. 1 ppm of a population is about 4.75 standard deviations away from the mean dose. For reference, Nvidia reported in 2011 that via defect levels need to be less than one in a billion, so that the minimum dose estimates above would need to be even tighter. 1 ppb of a population is 6 standard deviations away from the mean dose, raising the minimum dose requirement by 60%.

Ref.: SPIE Proc. 8326-96, 8683-36, 8679-50 (2013)

The wafer-per-hour (WPH) throughput estimates provided by EUV suppliers assume a dose of 15 mJ/cm², which currently allows about a minute of exposure time per wafer. However, the consideration of shot noise challenges this assumption, in which case EUV throughput targets can be put at risk, e.g., 60 WPH @15 mJ/cm² becomes 20 WPH @45 mJ/cm².

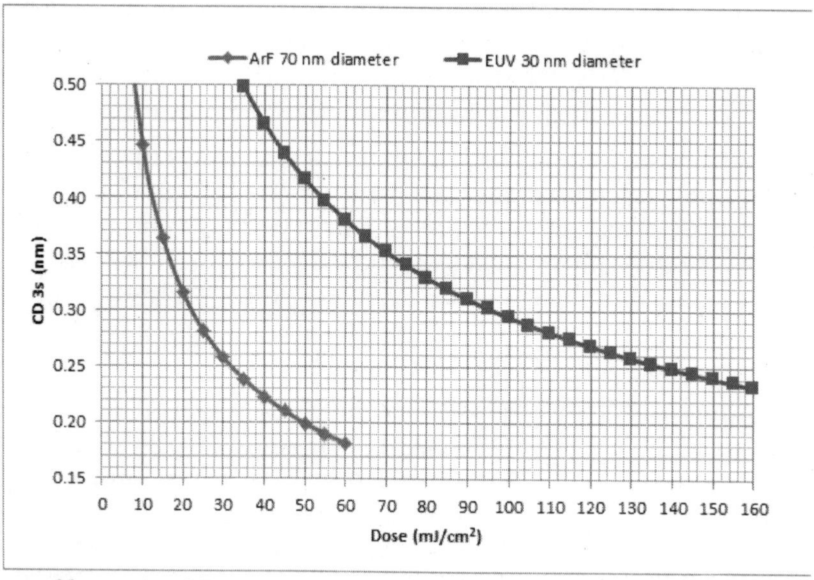

***Figure:** Shot noise CD 3s non-uniformity vs. exposure dose To match the shot noise performance for 193 nm 70 nm hole size at 25-35 mJ/cm², EUV exposure for 30 nm hole size must approach 160 mJ/cm². (Ref.: Z-Y. Pan* et al., *Proc. SPIE vol. 6924, 69241K (2008))*

TSMC also found that to match the shot noise performance for 193 nm light exposure at 70 nm hole size at 25-35 mJ/cm², the required dose for EUV exposure for 30 nm hole size must be more than 4 times larger, while to scale down the CD uniformity proportionately, it would have to be more than 16 times larger.

The shot noise issue is also applicable to the features patterned on masks used for EUV, targeted at 20 nm and below. A 12 uC/cm^2 absorbed dose used to pattern 80 nm contact holes on a mask (to print 20 nm on wafer) inevitably experiences 10% shot noise in the dose level over the population of a billion such contact holes.

The partially coherent light source is often represented as a collection of hundreds to thousands of points, each an independent source of photons. Furthermore, the asymmetric variation of the multilayer reflectivity with respect to different angles of incidence results in source points on one side being effectively brighter than those on the other. A million photons, e.g., 100 source points x 10,000 photons/point, at a dose of 10 photons/nm^2, would cover a 100,000 nm^2 area (~300 nm x 300 nm), far exceeding the theoretical resolution.

Carl Zeiss, the maker of the EUV Aerial Image Metrology System (AIMS), recently concluded that 15,000 photons per 18 nm pixel (corresponding to a dose of 68 mJ/cm^2) were required to guarantee sufficient CD fidelity.

The shot noise has strong bearing on the EUV source power issue mentioned above. For 10 mJ/cm^2, the power at intermediate focus should be 180 W; currently it is about 20 W at high duty cycle. However, significant shot noise may force minimum doses to be at least 42 mJ/cm^2 for 20 nm feature size (e.g., 20 nm cuts in 20 nm half-pitch lines) and 169 mJ/cm^2 for 10 nm feature size (e.g., 10 nm contacts on 14 nm half-pitch lines), therefore indicating the EUV source power to be a moving target becoming ever more difficult to reach. These minimum dose values already exceed the multilayer pulse damage thresholds indicated above. Actually, the most widely acknowledged concern of such high doses is the increased resist ougassing (30 mJ/cm^2 being prohibitive). Furthermore, if the doses increase by at least a factor of ~3, the crosslinking of the resist polymer becomes significant. As discussed below, due to high absorption, heating is more significant. For chemically amplified resists, higher dose exposure also increases line edge roughness due to acid generator decomposition. There could be some shot noise relief for the brighfield exposures which would be used for contact hole patterns with negative-tone metal oxide resists; flare has more severe impact (loss of image contrast) in brightfield exposures with higher doses. Soft x-ray exposures of HSQ resist have shown 50-70 nm linewidth increase related to increased reactions beyond exposure boundaries, due to dose increase in the 100 mJ/cm^2 range.

Proximity Effect (Secondary Electrons)

It is now recognised that for insulating materials like PMMA, low energy electrons can travel quite far (several nanometres is possible). For example, in sub-10 nm thick SiO_2, negligible electron scattering is expected. This is due to the fact that below the ionization potential the only energy loss mechanism is mainly through phonons and polarons. It should be noted that polaronic effects are manifest more strongly in ionic crystals than polymers and covalently bonded materials. In fact, polaron hopping could extend as far as 20 nm.

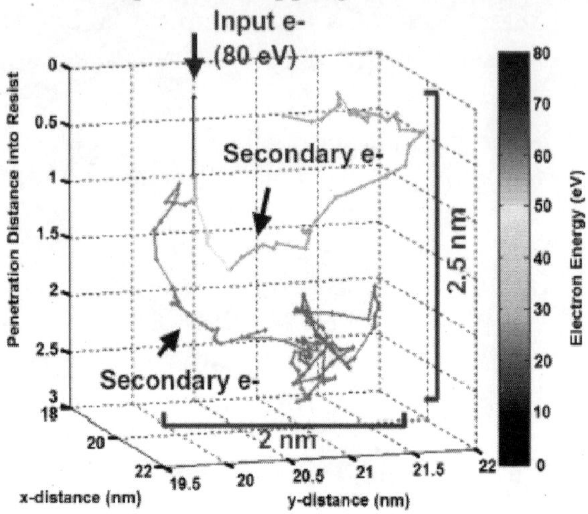

Figure: *Path of travel of secondary electrons. An 80 eV photoelectron generates a low energy secondary electron, which moves randomly within a resist volume (~2 nm in this case). Source: J. Torok* et al.*, J. Photopolym. Sci. Tech. vol. 26, 625 (2013).*

Recent studies indicate that the EUV secondary electron range in commercial resist is practically in the range of a few nanometres. This range has to be negligible (<10%) compared to the critical dimension, indicating current difficulties for sub-20 nm application.

Material	***10 eV Electron Inelastic Mean Free Path****
Water	10 nm
DNA	5 nm
PMMA	5 nm
SiO_2	7 nm

(*) On average, an electron with 10 eV energy travels this distance in the material before losing energy.

Material	***<3 eV Electron Attenuation Length***
Pentacene	7.5 ± 1.0 nm
Perylene	80 ± 8.0 nm

In a classic experiment by Feder et al. at IBM, an erbium layer on a PMMA resist layer was exposed to X-rays. The erbium layer absorbed the X-rays strongly, producing low energy secondary electrons. The X-rays which were not absorbed continued to penetrate into the PMMA, where they were only lightly absorbed. Upon removal of the erbium layer and subsequent PMMA development in solvent, the resist removal rate was found to be accelerated for the top 40 nm of the PMMA film, while it was much more gradual for the rest of the film. The accelerated rate was due to the secondary electron exposure, while the gradual rate was due to the X-ray absorption. This proved the maximum secondary electron exposure range of 40 nm in this case.

K. Murata also calculated the impact of 92 eV Auger electrons emitted into a layer of PMMA from a Si substrate during X-ray exposure. The range of exposure of the PMMA was 50 nm. A more recent experiment was performed by Carter et al. at MIT and University of Wisconsin–Madison, where the X-ray absorber generating the electrons was beneath the PMMA resist rather than on top of it. In this case, the accelerated dissolution of PMMA started approximately 50 nm above the substrate.

The significance of this secondary electron range is the appearance of a "proximity effect" for distances on the order of 50 nm or less. This causes the exposure tolerance to be reduced dramatically as feature sizes decrease below this range. Even though features can still print below this range, the resolution is affected by the randomness of energy distribution. The difference in experimentally determined ranges above (40 nm vs. 50 nm) is an indication of this fundamental variability. The secondary electron exposure can also be thought of as a blur effect. The blur is generally not included in optical-only image simulations. The proximity effect is also manifest by photoelectrons and secondary electrons leaving the top surface of the resist and then returning some tens of nanometres distance away. This also can be understood in terms of the emitted electrons forming a space charge cloud above the surface which is attracted to the positively charged surface in the vertical direction but laterally disperses (in vacuum) due to the negative charge mutual repulsion.

The secondary electron proximity effect was recently demonstrated by Stanford University using a scanning probe tip that emitted electrons in the 40–60 eV energy range. Dose sensitivity was demonstrated more than 25 nm away from the exposure centre. It indicates that within a 50 nm range of exposure widths, the low-energy (EUV-generated) electron distribution influences the linewidth distribution. This is a new effect not seen with conventional optical lithography. Photoelectron emission microscopy (PEEM) data was used to show that low energy electrons ~1.35 eV could travel as far as ~15 nm in SiO_2, despite an average measured attenuation length of 1.18 nm.

A study by the College of Nanoscale Science and Engineering (CNSE) presented at the 2013 EUVL Workshop indicated that, as a measure of EUV photoelectron and secondary electron blur, 50-100 eV electrons easily penetrated beyond 10 nm of resist thickness (PMMA or commercial resist); furthermore dielectric breakdown discharge is possible.

A 2012 study by Synopsys and IMEC revealed that the secondary electron effect on acid generation is on the order of several nm away from the initial secondary electron generation site. In combination with shot noise and post-exposure effects, this resulted in CDs ranging from 17 to 40 nm for a 32 nm half-pitch contact hole array.

Photoelectron Trajectories

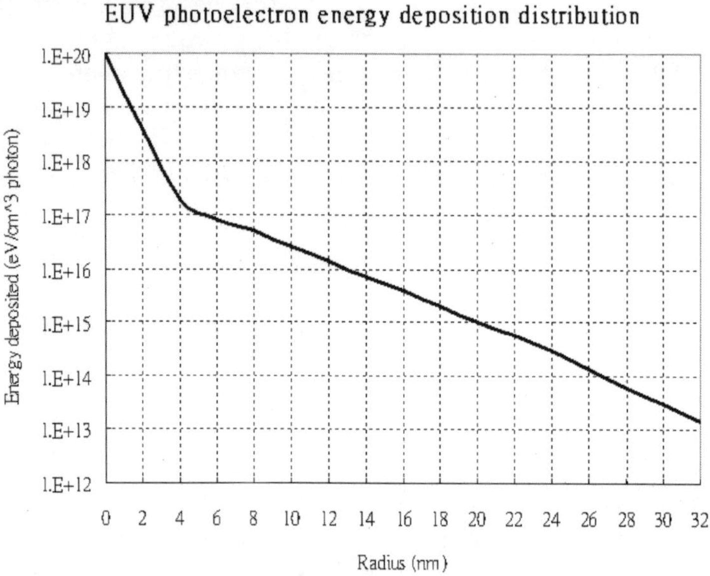

Figure: *Radial energy distribution from EUV photoelectrons. For photons incident at r=0, the radial range can be as large as 30 nm, but the highest amount of energy is deposited within a few nm distance. By adding up the radial energy distributions from many photons incident at different locations within a target area, the amount of deposited energy accumulates, enabling effects at further distances from the target area.*

Kotera et al. performed EUV photoelectron trajectory simulations, showing their range to be 30 nm. The spread of the energy deposition by these electrons can account for the observed line edge roughness. The top layer exposure is effectively less because electrons emitted from the surface never come back.

Efficient Photoresist Heating

Ritucci et al., reported on the improved thermal ablation efficiency for EUV wavelengths compared to DUV wavelengths. Since EUV exceeds the bandgap of all materials, it is more easily absorbed than longer wavelengths, and the same dose of incident energy results in more heating; even ~100 mJ/cm^2 would be hot enough to result in ablation. The resolution of chemically amplified photoresists is determined by thermally driven acid diffusion (spreading). It is worth noting that even at the ablation dose of 100 mJ/cm^2, the shot noise for a 1 nm pixel is still significant (3σ/avg = 36%), which could severely impact a critical dimension (CD) for which the pixel is at least 5%, i.e., 20 nm or less.

Point Spread Function of Resist

Kozawa et al. determined the point spread function of EUV chemically amplified resists using a basic acid generation calculation and simulation fit. The range of acid generation extended ~20 nm from the absorption point, entailing a ~40 nm resolution limit. Given that photoresists easily diffuse acid molecules, it would be no surprise that the smaller and lighter electrons produced by EUV or other ionizing radiation would diffuse faster and further, rendering the expected optical resolution meaningless.

The resist blur based on print results at the end of 2008 is in the range of 10-16 nm. Half-pitch resolution is still a struggle below 30 nm, and line edge roughness is still a major issue. A study in 2011 focusing on 22 nm and 24 nm half-pitch indicated a temperature-dependent blur for the post-exposure bake process, ranging from ~5 nm at 80 °C to ~10 nm at ~110 °C. The secondary electron blur was reported to be not observed in this range. The aerial images were corrected for the estimated flare (which would include any long-range secondary electron blur).

Uncertainty Principle

As secondary electron generation involves inelastic scattering with momentum transfer, there will be an associated position uncertainty. As lower energy electrons have less momentum transfer, the delocalization of the secondary electron generation process tends to be higher (~nm), which would have a more direct impact on LER.

EUV Diffraction and Multilayer Angle Bandwidth

Resolution requirements below 20 nm require higher numerical aperture (N.A.) EUV optical systems supported by multilayer optics. However, the larger aperture necessarily entails larger angles of incidence as well as a larger range of incident angles. The multilayer currently used in EUV masks and optical systems tends to cut off light at larger angles which is required to image tighter pitches. For 2D patterns like dense contact holes, contrast already decreases from 80% to <75% going from 19 nm to 18 nm half-pitch, 14 nm half-pitch is already in need of multilayer tuning, and in fact, without the use of increased demagnification (above 4X) to help reduce the angles, no imaging beyond ~13 nm half-pitch is possible. Such demagnification would result in large photomask substrate sizes or else multiple small fields (on different masks).

In fact, for EUV mask pitches of 8 wavelengths or less (demagnified 4x to 2 wavelengths (13-14 nm half-pitch) or less on the wafer), diffraction into the multilayer at larger angles is another source of significant image degradation, which requires intensive computation to evaluate. For larger angles, the multilayer reflectance decreases significantly. Rigorous EM simulations for the EUV binary mask at different magnifications have already indicated that for the standard 4X magnification, the diffraction order efficiencies starts varying significantly with new asymmetry below 20 nm half-pitch.

EUV Mask Absorber Thickness Variation

A study published in 2011 by ASML, Brion and Zeiss found that EUV mask absorber thickness has significant effect on exposure latitude and mask error enhancement. This was observed at 27-32 nm half-pitch. Furthermore, the effects were noticeably different for different conventional illumination settings. For example, for 32 nm line-and-space patterns, the optimum dose-to-size differed by more than 30%, while best bias differed by more than 4 nm, between 55.4 nm and 58 nm thickness (+/- 2.3% thickness variation). The mask absorber thickness could be optimized for size and exposure latitude for different conventional mask patterns. There remains a tradeoff of exposure latitude vs. throughput. The absorber still allows some fraction of light to pass in one direction, and the top surface itself reflects some light. The optimum absorber thickness is a linear function of the half-pitch, as well as the range of incident angles. Even at the optimum absorber thickness, there is sufficient reflectivity to present a field-to-field flare effect from the mask border. The shadowing effect is more obvious with thicker absorber. The absorber thickness needs to be tightly controlled across the mask and from mask-to-mask to prevent proximity matching errors. Furthermore, the absorber thickness needs to be considered for OPC determination. With a typical CD sensitity of around 1 nm/cm^{2,} for 27-32 nm lines, the issue is severe enough that there is a proposal to completely etch away the absorber and multilayer at the mask image field border, which entails a second layer or level of patterning in the mask fabrication, as in the case of a phase-shift mask.

Higher Magnification Mask Stitching

For higher numerical apertures, as noted above, it would be necessary to increase the angle of reflection in the EUV optical system. However, this could require returning of the multilayers. As an alternative, the magnification of the system may be increased beyond 4X, for example to 8X. However, this would require either larger EUV mask substrates, or else the division of chip patterns among two or more conventional 6– EUV masks. The chip pattern would have to be constructed from stitching two or more likely four sub-patterns together. In this case, the mask change time is a key influence on the overall throughput.

EUVL Demonstrations

Interference lithography at the Paul Scherrer Institute has been used to demonstrate sub-10 nm line-space features. The resist performance tested with this source does not reflect the performance expected for an EUV projection tool due to the limited contrast of projection tools.

In 1996, a collaboration between Sandia National Laboratories, University of California at Berkeley, and Lucent Technologies, produced NMOS transistors with gate lengths from 75 nm to 180 nm. The gate lengths were defined by EUV lithography. The device saturation current at 130 nm gate length was ~0.2 mA/um. A 100 nm gate device showed subthreshold swing of 90 mV/decade and saturated transconductance of 250 mS/mm. A commercial NMOS at the same design rule patterned by then-state-of-the-art DUV lithography showed 0.94 mA/um saturation current and 860 mS/mm saturated transconductance. The subthreshold swing in this case was less than 90 mV/decade.

In February 2008, a collaboration including IBM and AMD, based at the College of Nanoscale Science and Engineering (CNSE) in Albany, New York, used EUV lithography to pattern 90 nm trenches in the first metal layer of a 45 nm node test chip. No specific details on device performance were given. However, the lithographic performance details given still indicated much to be desired:

- CD uniformity: 6.6%
- Overlay: 17.9 nm x, 15.6 nm y, possibly correctable to 6.7 nm x, 5.9 nm y
- Power: 1 W at wafer (>200 W required for high volume), with a dose of 3.75 mJ/cm^2
- Defects: 1/sq. cm.

The high defect level may not be unexpected as AMD's 45 nm node Metal 1 design rule was 90 nm while the same EUV exposure theoretically could result in printed defects below 30 nm originating from mask defects larger than 100 nm. Optical lithography pushed beyond its natural resolution limit has a significant advantage in this regard.

Apparently, the CNSE EUV tool suffered from a well-known 16% flare problem. Flare effects may be difficult to separate from the secondary electron effects discussed earlier.

Also in July 2008, IMEC printed ~60 nm contacts using their installed EUV tool. Doses of 12–18 mJ/cm^2 were used.

In August 2008, SEMATECH demonstrated a 22 nm half-pitch using chemically-amplified photoresist. However, even at 15 mJ/cm^2, the linewidth roughness was very significant, 5–6 nm, so that even the image pitch regularity was challenged.

In April 2009, IMEC fabricated 22 nm SRAM cells where the contact and Metal 1 layers (~45 nm design rule) were printed with EUV lithography. However, it was acknowledged that EUV would not be ready when companies start using 22 nm. In addition, it was commented that the feature edge profiles indicated slope asymmetry related to the characteristic EUV illumination asymmetry. Whereas this demonstration only focused on a limited number of ~45 nm features, Intel's shot noise calculation above for billions of features ~30 nm indicates difficult challenges ahead for manufacturing.

In late 2009, KLA-Tencor and GlobalFoundries along with Lawrence Berkeley National Labs published a paper which showed the stochastic behaviour of EUV-generated secondary electrons in EUV resists. In particular, 32 nm half-pitch trenches showed significant edge roughness, width roughness and critical dimension (CD) variability. It may also explain the ~ 15 nm resist blur observed in an earlier study.

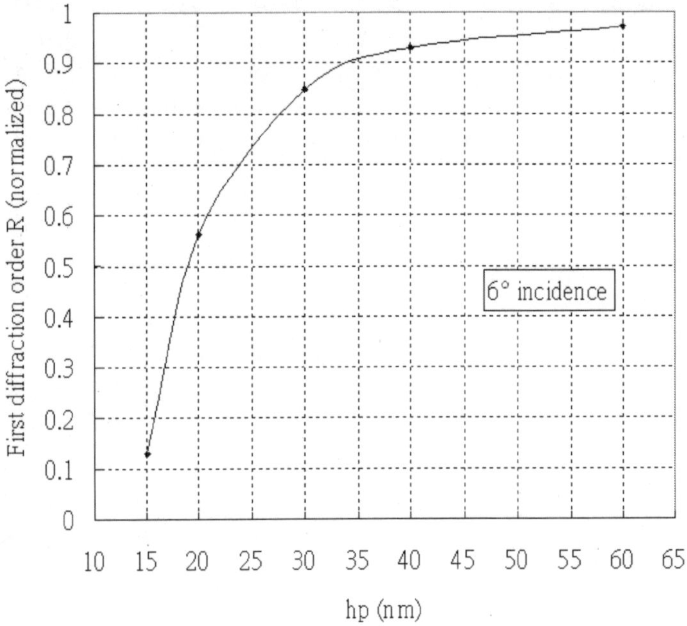

Figure: *Multilayer effect on pitch resolution. As the pitch decreases, the diffraction angle of the first order increases, leading to less pattern information in the EUV light reflected by the mask multilayer to be collected by the optics. This leads to an additional source of resolution loss.*

EUV Device Damage

MOSFETs made with Ge showed large sensitivity to EUV doses, starting to degrade even at levels of 10-20 mJ/cm^2.

Timing Impact

As early as 2003, there was still optimism that EUVL would quickly be commercialized, as exemplified by the following quote: "Good progress has been made on all the technology fronts. Commercialization in 2009 remains the main goal for all the EUV community." However, as of 2012, EUVL still has not been commercialized.

The difficulties of EUV stem mainly from the difficulty of EUV light generation and the sensitivity of EUV light to surface contamination and roughness. The rise of multiple patterning, essentially a wavelength-independent technique, is also further jeopardizing its introduction. Intel's Justin Rattner has said that creation of the needed (longer wavelength) masks was a prime use of high performance computing.

2009

In July 2009, ASML Holding and Cymer Inc. announced the shipment of the first fully integrated laser-produced plasma (LPP) extreme ultraviolet (EUV) lithography light source from Cymer to ASML, where it is supporting integration and testing of next generation EUV lithography scanners.

In July 2009, a review of the EUV infrastructure indicated that there were significant gaps remaining, in particular for mask defect inspection. The gaps may not be filled until 2013 or later, which would impact the rollout of EUV for high-volume manufacturing by 2015.

It was most recently indicated that EUV mask features ~100 nm half-pitch (on the mask itself) were at the current resolution limit of EUV mask generation. Since these features would be demagnified 4X for upcoming 300 mm tools, it would indicate the mask itself would not be able to conventionally print good images at ~25 nm half-pitch or below. The same study reiterated that resolution was not limited by the optics but by the resist response, which prevented sub-20 nm half-pitch even with mask resolution enhancement. This resolution limit on the mask itself could be tied to the shot noise issue pointed out above which is involved in the fabrication of the mask.

2010

In June 2010, TSMC announced it would take delivery of an EUV system for R&D purposes, to be delivered in 2011. The tool's weight reportedly warranted purchasing a special clamp costing more than 2 million USD.

In July 2010, at Semicon West 2010, ASML updated the resolution achieved by EUVL, the status being 24 nm in 2010, but with a severely degraded NILS (normalized image log-slope) below 1.0, indicating poor lithographic quality. In fact, as finer resolution is projected, the NILS is expected to go to 0, indicating impossibility of EUVL use.

In 2010, Samsung cited various additional processes still needed to be developed to address line-width roughness (LWR), arising from the shot noise discussed earlier, which would drive up costs. In addition, it cited out-of-band wavelengths in the DUV range as causing flare effects; this was also observed by others.

At SPIE 2010, researchers from the Centre for X-ray Optics at Lawrence Berkeley National Laboratory and the University of California, Berkeley presented results showing that replicated surface roughness (RSR) from the substrate to the multilayer surface as low as 50 pm could induce phase errors that are imaged as speckle with defocus, which would lead to line-edge roughness (LER) that could not be tolerated for the 22 nm and 16 nm nodes. This spec is much smaller than current capabilities. For reference, the Bohr radius is 53 pm and the rms surface roughness of silicon wafers is on the order of 100 pm.

ASML EUV Tools in Use or in Development

Available Year	*EUV Tool*	*Resolution*	*Throughput goal*	*Dose, Source*
2006	ADT	32 nm	4 WPH	5 mJ/cm^2, ~8W
2010	NXE:3100	27 nm	60 WPH	10 mJ/cm^2, >100W
2012	NXE:3300B	22 nm	125 WPH	15 mJ/cm^2, >250W
2013	NXE:3300C	16 nm if <7 nm resist diffusion length	150 WPH	15 mJ/cm^2, >350W

Source: ASML, International Workshop on EUVL, Maui 2010

NXE:3300B stepper uses 13.5 nm EUV light generated by very hot tin. High-energy laser turns microscopic droplets of molten tin into a plasma which emits EUV light.

Timing for 1x nm and Beyond

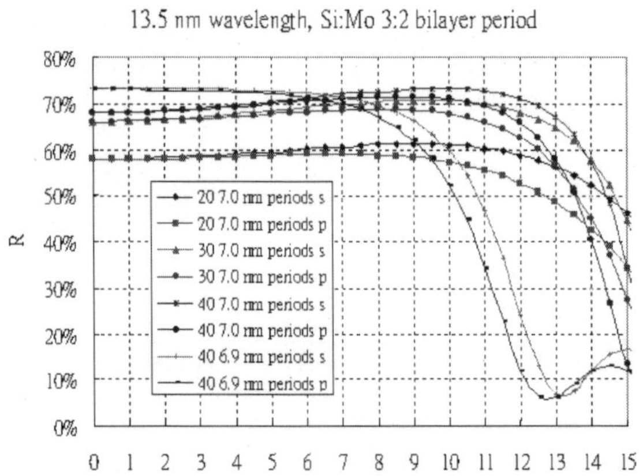

Figure: *Polarization dependence of multilayer reflectivity. The s and p polarizations are reflected differently by the multilayers at larger angles.*

Hynix has found that 193 nm immersion lithography with double patterning could resolve down to ~ 20 nm, while the more cost-conscious EUV approach could only resolve down to 28 nm, less than a generation's improvement over 193 nm immersion lithography single patterning's resolution limit of 38 nm.

Due to its current resolution limitations and lack of readiness, EUVL has recently been supplanted by multiple patterning using existing immersion lithography tools as the lithography of choice for current 20-nm class NAND Flash technology generations and beyond. Samsung and IM Flash Technologies have already started using spacer double patterning for their 3X nm and 2X nm NAND Flash. Toshiba and SanDisk are already starting to make 19 nm NAND flash in 2011. In fact, with EUV single exposure resolution still exceeding 22 nm, multiple patterning would be necessary for ~10-14 nm regardless of using EUV or ArF wavelength. At IEDM 2011, Hynix is expected to report on 1X nm NAND technology using quadruple spacer patterning.

In September 2009, Intel disclosed that for its 15 nm process, EUV did not appear to be ready in timely fashion. Hence, Intel is preparing to extend 193 nm immersion lithography with double and possibly triple patterning to 15 nm. Currently, it is also exploring the feasibility of EUV for its 11 nm logic node; however, 2D contact patterning still appears only good to ~ 26 nm half-pitch, while a special rinse for improving line width roughness appears very pitch-dependent. Intel has declared that even without EUV, it could still use 193 nm immersion lithography at the 11 nm node.

As of 2010, it has been observed that EUV would be delayed until 2015 to allow the development of the required EUV inspection tools. On the other hand, manufacturer roadmaps show 2012–2013 as the timing needed for 1x nm, precluding the use of EUV. Consideration of long-term implementation of EUVL will now need to consider sub-10 nm design rules. However, at this point, due to larger electric field vector angles, polarization effects become important. The effective flare for 10 nm half-pitch lines-spaces due to polarization effects is 11%. The difficulty of EUV is the lack of polarization control (no laser sources, only plasmas).

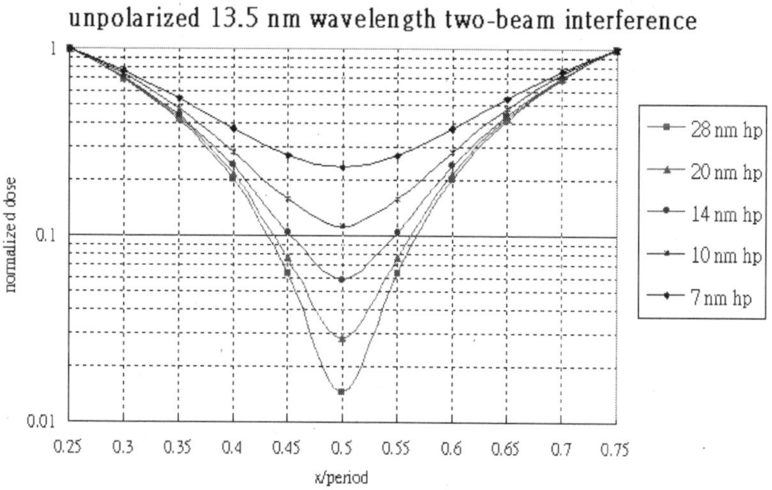

Figure: *Polarization effects. As pitch is reduced, the image is more affected by polarization, resulting in reduced contrast in the unpolarized image. At 14 nm hp, for example, the minimum intensity is already ~6% of the median value.*

ASML has indicated that below 16 nm capability would require a larger illumination angle than the current 6°, and hence require a fundamental change to the current EUV multilayer optics infrastructure. In particular, apodization (non-uniformity of intensity across the light entrance pupil), due to different reflectivity at different angles, becomes more severe for higher numerical apertures.

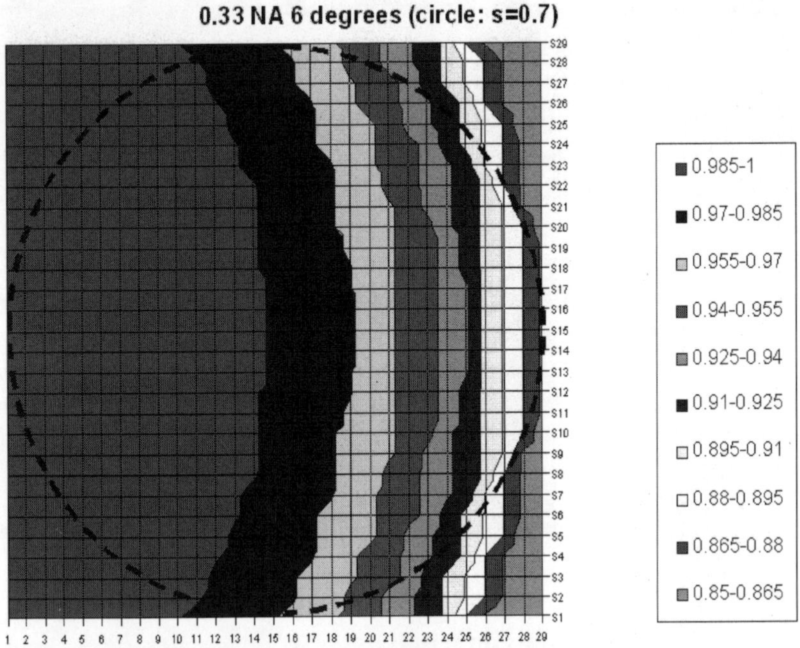

Figure: *Apodization by EUV multilayer The angle-dependent reflectivity of the EUV multilayer results in non-uniform intensity across the pupil. Imaging can be significantly affected for differences of 10% or more.*

Canon indicated at the 1st International EUVL Symposium that higher NA EUV systems would exhibit a larger difference of system transmittance between the two different polarizations, due to the larger range of angles being accessed. Another issue is that for sub-10 nm applications, the electron beam lithography for EUV mask patterning, already burdened by throughput issues, will have practical resist difficulties for meeting the resolution requirement. In fact, for 20 nm and below, the current electron-beam mask writers cannot repeatably deliver 80 nm sizes on the mask, which corresponds to 20 nm on the water.

Recent EUV zone plate lithography reports have indicated that for sub-20 nm half-pitch, even next-generation lithography would require double patterning; in fact, double patterning may not be sufficient even for sub-12 nm half-pitch. As a result, the overlay and throughput requirements would need to be at least twice as stringent as before.

At the 2011 LithoVision conference, Intel indicated that EUV technology is already late for even Intel's 10 nm design rule planning.

ASML has suggested that the 13.5 nm EUV wavelength is expected to be used down to 10 nm, beyond which a new EUV wavelength of 6.6 to 6.8 nm is expected to be used for finer resolution. Tools currently projected through 2015 are not expected to reach 15 nm resolution.

Cymer delayed its EUV 20 W source delivery from first quarter to second quarter of 2012.

Samsung also delayed its EUV to after 2013. Likewise, Global Foundries and TSMC are delaying EUV use to beyond 20 nm node.

As of beginning of 2012, EUV has significant issues remaining in the areas of source power, defects, overlay, resist, and mask.

As of July 2012, 6 NXE:3100 tools (now discontinued) and 11 NXE:3300 tools have been ordered for process development and 4 NXE:3300 tools targeted for production; the latter order was split between Samsung and SKHynix.

At the 2013 EUVL Workshop, Intel announced that EUV would still be under development in 2015, and hence would be targeted for 2017 7 nm HVM. Consequently, 10 nm would be carried out with ArF immersion multiple patterning. Moreover, ArF immersion pitch quartering seems more likely than EUV (single patterning) to be extended to 7 nm by avoiding going below the k1=0.3 limit. TSMC and GlobalFoundries have made similar statements.

At the 2013 iEUVi Mask TWG update, it was revealed that the EUV mask infrastructure will have to be re-defined to allow the scaling to 14 nm half-pitch and beyond. This roadmap clarification also means continued delays for EUV.

At its Q4 2013 earnings conference call in January 2014, ASML mentioned that 19 nm node DRAM would be using double patterning with DUV rather than EUV and that EUV development by ASML with customers for 16 nm DRAM would start later in the year.

As of January 2014, the EUV production throughput capability is 50 WPH; its next line of EUV tools, the NXE:3350, is planned to have EUV source self-cleaning capability. As contingency for EUV, ASML is already shipping the NXT:1970Ci, an immersion tool capable of multiple patterning.

At the 2013 EUVL Workshop, a collaboration (Applied Materials/IMEC/Micron) study on the implementation of spacer patterning for EUVL indicated that the local CD uniformity of ~30 nm contact holes was influenced by shot noise. Furthermore, it was indicated that the current choice for low 1xnm half-pitch patterning is 193i SAQP (already being used for 16 nm NAND by Micron) or EUV SADP.

Resource Requirements: EUV vs. ArF Immersion Double Patterning

Utility	*200 W output EUV*	*90 W output ArF immersion double patterning*
Electrical power (kW)	532	49
Cooling water flow (L/min)	1600	75
Gas lines	6	3

Source: Gigaphoton, Sematech Symposium Japan, September 15, 2010

The required utility resources are significantly larger for EUV compared to 193 nm immersion, even with double patterning. Hynix reported at the 2009 EUV Symposium that the wall plug efficiency was ~0.02% for EUV, i.e., to get 200 W at intermediate focus for 100 WPH, one would require 1MW of input power, compared to 165 kW for an ArF immersion scanner, and that even at the same throughput, the footprint of the EUV scanner was ~3x the footprint of an ArF immersion scanner, resulting in productivity loss. Additionally, to confine ion debris, a superconducting magnet may be required.

EUV with Double Patterning

The resolution limit for EUV single patterning is ~15-19 nm half-pitch, while that for ArF double patterning is 30-20 nm half-pitch, which suggests EUV double patterning rather than single patterning should succeed ArF double patterning. Owing to the various current limitations on EUV reaching 10 nm half-pitch, such as the current EUV multilayer angle bandwidth for ~13-14 nm half-pitch, double patterning is the planned approach to extend planned 0.33 NA NXE:3300 tools to 10 nm half-pitch. The currently expected year of introduction of EUV into production is 2015 at roughly 20 nm half-pitch, allowing one generation (~20-13 nm) of single-exposure patterning use. In Intel's complementary lithography scheme at 20 nm half-pitch, EUV would be used only in a second line-cutting exposure after a first 193 nm line-printing exposure. The cost may be compared to the expected multiple exposures using 193 nm light only. Some ~20 nm half-pitch patterns, e.g., 22 nm half-pitch DRAM active areas, may be patterned by a single 193 nm exposure using a special mask. At the ITRS 2012 Winter Conference, the Lithography Update indicated the use of EUV with double patterning for 14 nm half-pitch in 2017. A return to extended generations of single exposure patterning would be possible with a wavelength even shorter than the 13.5 nm EUV wavelength. Such a wavelength (~6.7 nm) would be beyond EUV, and is often referred to as BEUV (Beyond Extreme UltraViolet).

Electron Beam Lithography

Electron-beam lithography (often abbreviated as e-beam lithography) is the practice of scanning a focused beam of electrons to draw custom shapes on a surface covered with an electron sensitive film called a resist ("exposing"),. The electron beam changes the solubility of the resist enabling selective removal of either the exposed or non-exposed regions of the resist by immersing it in a solvent ("developing"). The purpose, as with photolithography, is to create very small structures in the resist that can subsequently be transferred to the substrate material, often by etching.

The primary advantage of electron-beam lithography is that it can draw custom patterns (direct-write) with sub-10 nm resolution. This form of maskless lithography has high resolution and low throughput, limiting its usage to photomask fabrication, low-volume production of semiconductor devices, and research & development.

Electron-Beam Lithography Systems

Electron-beam lithography systems used in commercial applications are dedicated e-beam writing systems that are very expensive (> US$1M). For research applications, it is very common to convert an electron microscope into an electron beam lithography system using a relatively low cost accessories (< US$100K). Such converted systems have produced linewidths of ~20 nm since at least 1990, while current dedicated systems have produced linewidths on the order of 10 nm or smaller. Electron-beam lithography systems can be classified according to both beam shape and beam deflection strategy. Older systems used Gaussian-shaped beams and scanned these beams in a raster fashion. Newer systems use shaped beams, which may be deflected to various positions in the writing field (this is also known as vector scan).

Electron Sources

Lower-resolution systems can use thermionic sources, which are usually formed from lanthanum hexaboride. However, systems with higher-resolution requirements need to use field electron emission sources, such as heated W/ZrO_2 for lower energy spread and enhanced brightness. Thermal field emission sources are preferred over cold emission sources, in spite of the former's slightly larger beam size, because they offer better stability over typical writing times of several hours.

Lenses

Both electrostatic and magnetic lenses may be used. However, electrostatic lenses have more aberrations and so are not used for fine focusing. There is no current mechanism to make achromatic electron beam lenses, so extremely narrow dispersions of the electron beam energy are needed for finest focusing.

Stage, Stitching and Alignment

Typically, for very small beam deflections electrostatic deflection "lenses" are used, larger beam deflections require electromagnetic scanning. Because of the inaccuracy and because of the finite number of steps in the exposure grid the writing field is of the order of 100 micrometre – 1 mm. Larger patterns require stage moves. An accurate stage is critical for stitching (tiling writing fields exactly against each other) and pattern overlay (aligning a pattern to a previously made one).

Electron Beam Write Time

The minimum time to expose a given area for a given dose is given by the following formula:

$$D \cdot A = T \cdot I$$

where T is the time to expose the object (can be divided into exposure time/step size), I is the beam current, D is the dose and A is the area exposed.

For example, assuming an exposure area of 1 cm^2, a dose of 10^{-3} coulombs/cm^2, and a beam current of 10^{-9} amperes, the resulting minimum write time would be 10^6 seconds (about 12 days). This minimum write time does not include time for the stage to move back and forth, as well as time for the beam to be blanked (blocked from the wafer during deflection), as well as time for other possible beam corrections and adjustments in the middle of writing. To cover the 700 cm^2 surface area of a 300 mm silicon wafer, the minimum write time would extend to $7*10^8$ seconds, about 22 years. This is a factor of about 10 million times slower than current optical lithography tools. It is clear that throughput is a serious limitation for electron beam lithography, especially when writing dense patterns over a large area.

E-beam lithography is not suitable for high-volume manufacturing because of its limited throughput. The smaller field of electron beam writing makes for very slow pattern generation compared with photolithography (the current standard) because more exposure fields must be scanned to form the final pattern area ($\leq mm^2$ for electron beam vs. ≥ 40 mm^2 for an optical mask projection scanner). The stage moves in between field scans. The electron beam field is small enough that a rastering or serpentine stage motion is needed to pattern a 26 mm X 33 mm area for example, whereas in a photolithography scanner only a one-dimensional motion of a 26 mm X 2 mm slit field would be required.

Currently an optical maskless lithography tool is much faster than an electron beam tool used at the same resolution for photomask patterning.

Shot Noise

As features sizes shrink, the number of incident electrons at fixed dose also shrinks. As soon as the number reaches ~10000, shot noise effects become predominant, leading to substantial natural dose variation within a large feature population. With each successive process node, as the feature area is halved, the minimum dose must double to maintain the same noise level. Consequently, the tool throughput would be halved with each successive process node.

Feature Diameter (nm)	*Minimum Dose for One-in -a-million 5% dose error ($\mu C/cm^2$)*
40	127
28	260
20	509
14	1039
10	2037
7	4158

Note: 1 ppm of population is about 5 standard deviations away from the mean dose.

Ref.: SPIE Proc. 8683-36 (2013)

Shot noise is a significant consideration even for mask fabrication. For example, a commercial mask e-beam resist like FEP-171 would use doses less than 10 $\mu C/cm^2$, whereas this leads to noticeable shot noise for a target CD even on the order of ~200 nm on the mask.

Defects in Electron-Beam Lithography

Despite the high resolution of electron-beam lithography, the generation of defects during electron-beam lithography is often not considered by users. Defects may be classified into two categories: data-related defects, and physical defects. Data-related defects may be classified further into two sub-categories. Blanking or deflection errors occur when the electron beam is not deflected properly when it is supposed to, while shaping errors occur in variable-shaped beam systems when the wrong shape is projected onto the sample. These errors can originate either from the electron optical control hardware or the input data that was taped out. As might be expected, larger data files are more susceptible to data-related defects.

Physical defects are more varied, and can include sample charging (either negative or positive), backscattering calculation errors, dose errors, fogging (long-range reflection of backscattered electrons), outgassing, contamination, beam drift and particles. Since the write time for electron beam lithography can easily exceed a day, "randomly occurring" defects are more likely to occur. Here again, larger data files can present more opportunities for defects.

Photomask defects largely originate during the electron beam lithography used for pattern definition.

Electron Energy Deposition in Matter

The primary electrons in the incident beam lose energy upon entering a material through inelastic scattering or collisions with other electrons. In such a collision the momentum transfer from the incident electron to an atomic electron can be expressed as $dp = 2e^2 / bv$, where b is the distance of closest approach between the electrons, and v is the incident electron velocity.

The energy transferred by the collision is given by $T = (dp)^2 / 2m = e^4 / Eb^2$, where m is the electron mass and E is the incident electron energy, given by $E = (1/2)mv^2$.

By integrating over all values of T between the lowest binding energy, E_0 and the incident energy, one obtains the result that the total cross section for collision is inversely proportional to the incident energy E, and proportional to $1/E_0 - 1/E$. Generally, $E >> E_0$, so the result is essentially inversely proportional to the binding energy.

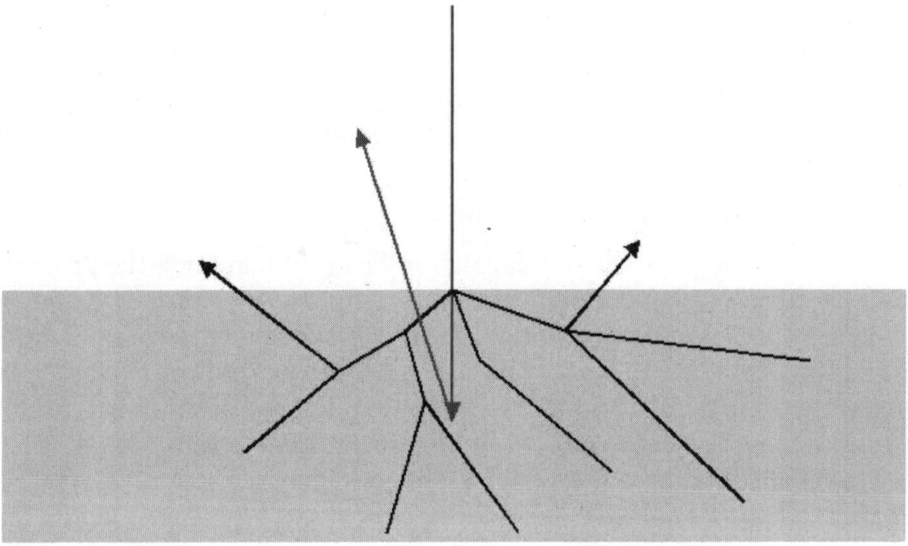

***Figure:** Electron trajectories in resist: An incident electron (purple) produces secondary electrons (blue). Sometimes, the incident electron may itself be backscattered as shown here and leave the surface of the resist (amber).*

By using the same integration approach, but over the range $2E_0$ to E, one obtains by comparing cross-sections that half of the inelastic collisions of the incident electrons produce electrons with kinetic energy greater than E_0. These secondary electrons are capable of breaking bonds (with binding energy E_0) at some distance away from the original collision. Additionally, they can generate additional, lower energy electrons, resulting in an electron cascade. Hence, it is important to recognise the significant contribution of secondary electrons to the spread of the energy deposition.

In general, for a molecule AB:

$$e^- + AB \rightarrow AB^- \rightarrow A + B^-$$

This reaction, also known as "electron attachment" or "dissociative electron attachment" is most likely to occur after the electron has essentially slowed to a halt, since it is easiest to capture at that point. The cross-section for electron attachment is inversely proportional to electron energy at high energies, but approaches a maximum limiting value at zero energy. On the other hand, it is already known that the mean free path at the lowest energies (few to several eV or less, where dissociative attachment is significant) is well over 10 nm, thus limiting the ability to consistently achieve resolution at this scale.

Resolution Capability

With today's electron optics, electron beam widths can routinely go down to a few nm. This is limited mainly by aberrations and space charge. However, the feature resolution limit is determined not by the beam size but by forward scattering (or effective beam broadening) in the resist while the pitch resolution limit is determined by secondary electron travel in the resist.

This point is driven home by the 2007 demonstration of double patterning using electron beam lithography in the fabrication of 15 nm half-pitch zone plates. Although a 15 nm feature was resolved, a 30 nm pitch was still difficult to do, due to secondary electrons scattering from the adjacent feature.

The use of double patterning allowed the spacing between features to be wide enough for the secondary electron scattering to be significantly reduced. The forward scattering can be decreased by using higher energy electrons or thinner resist, but the generation of secondary electrons is inevitable. It is now recognised that for insulating materials like PMMA, low energy electrons can travel quite a far distance (several nm is possible). This is due to the fact that below the ionization potential the only energy loss mechanism is mainly through phonons and polarons, although the latter is basically an ionic lattice effect. Polaron hopping could extend as far as 20 nm. The travel distance of secondary electrons is not a fundamentally derived physical value, but a statistical parameter often determined from many experiments or Monte Carlo simulations down to < 1 eV. This is necessary since the energy distribution of secondary electrons peaks well below 10 eV. Hence, the resolution limit is not usually cited as a well-fixed number as with an optical diffraction-limited system. Repeatability and control at the practical resolution limit often require considerations not related to image formation, e.g., resist development and intermolecular forces.

A study by the College of Nanoscale Science and Engineering (CNSE) presented at the 2013 EUVL Workshop indicated that, as a measure of electron blur, 50-100 eV electrons easily penetrated beyond 10 nm of resist thickness (PMMA or commercial resist); furthermore dielectric breakdown discharge is possible.

Scattering

In addition to producing secondary electrons, primary electrons from the incident beam with sufficient energy to penetrate the resist can be multiply scattered over large distances from underlying films and/or the substrate. This leads to exposure of areas at a significant distance from the desired exposure location.

For thicker electrons, as the primary electrons move forward, they have an increasing opportunity to scatter laterally from the beam-defined location. This scattering is called forward scattering. Sometimes the primary electrons are scattered at angles exceeding 90 degrees, i.e., they no longer advance further into the resist. These electrons are called backscattered electrons and have the same effect as long-range flare in optical projection systems. A large enough dose of backscattered electrons can lead to complete exposure of resist over an area much larger than defined by the beam spot.

Proximity Effect

The smallest features produced by electron-beam lithography have generally been isolated features, as nested features exacerbate the proximity effect, whereby electrons from exposure of an adjacent region spill over into the exposure of the currently written feature, effectively enlarging its image, and reducing its contrast, i.e., difference between maximum and minimum intensity. Hence, nested feature resolution is harder to control. For most resists, it is difficult to go below 25 nm lines and spaces, and a limit of 20 nm lines and spaces has been found. In actuality, though, the range of secondary electron scattering is quite far, sometimes exceeding 100 nm, but becoming very significant below 30 nm.

The proximity effect is also manifest by secondary electrons leaving the top surface of the resist and then returning some tens of nanometres distance away.

Proximity effects (due to electron scattering) can be addressed by solving the inverse problem and calculating the exposure function $E(x,y)$ that leads to a dose distribution as close as possible to the desired dose $D(x,y)$ when convolved by the scattering distribution point spread function $PSF(x,y)$. However, it must be remembered that an error in the applied dose (e.g., from shot noise) would cause the proximity effect correction to fail.

Charging

Since electrons are charged particles, they tend to charge the substrate negatively unless they can quickly gain access to a path to ground. For a high-energy beam incident on a silicon wafer, virtually all the electrons stop in the wafer where they can follow a path to ground. However, for a quartz substrate such as a photomask, the embedded electrons will take a much longer time to move to ground. Often the negative charge acquired by a substrate can be compensated or even exceeded by a positive charge on the surface due to secondary electron emission into the vacuum.

The presence of a thin conducting layer above or below the resist is generally of limited use for high energy (50 keV or more) electron beams, since most electrons pass through the layer into the substrate. The charge dissipation layer is generally useful only around or below 10 keV, since the resist is thinner and most of the electrons either stop in the resist or close to the conducting layer. However, they are of limited use due to their high sheet resistance, which can lead to ineffective grounding.

The range of low-energy secondary electrons (the largest component of the free electron population in the resist-substrate system) which can contribute to charging is not a fixed number but can vary from 0 to as high as 50 nm. Hence, resist-substrate charging is not repeatable and is difficult to compensate consistently. Negative charging deflects the electron beam away from the charged area while positive charging deflects the electron beam toward the charged area.

Electron-Beam Resist Performance

Due to the scission efficiency generally being an order of magnitude higher than the crosslinking efficiency, most polymers used for positive-tone electron-beam lithography will crosslink (and therefore become negative tone) at doses an order of magnitude than doses used for positive tone exposure. Such large dose increases may be required to avoid shot noise effects.

A study performed at the Naval Research Laboratory indicated that low-energy (10–50 eV) electrons were able to damage ~30 nm thick PMMA films. The damage was manifest as a loss of material.

- For the popular electron-beam resist ZEP-520, a pitch resolution limit of 60 nm (30 nm lines and spaces), independent of thickness and beam energy, was found.
- A 20 nm resolution had also been demonstrated using a 3 nm 100 keV electron beam and PMMA resist. 20 nm unexposed gaps between exposed lines showed inadvertent exposure by secondary electrons.
- Hydrogen silsesquioxane (HSQ) is a negative tone resist that is capable of forming isolated 2-nm-wide lines and 10 nm periodic dot arrays (10 nm pitch) in very thin layers. HSQ itself is similar to porous, hydrogenated SiO_2. It may be used to etch silicon but not silicon dioxide or other similar dielectrics.

New Frontiers in Electron-Beam Lithography

To get around the secondary electron generation, it will be imperative to use low-energy electrons as the primary radiation to expose resist. Ideally, these electrons should have energies on the order of not much more than several eV in order to expose the resist without generating any secondary electrons, since they will not have sufficient excess energy. Such exposure has been demonstrated using a scanning tunneling microscope as the electron beam source. The data suggest that electrons with energies as low as 12 eV can penetrate 50 nm thick polymer resist. The drawback to using low energy electrons is that it is hard to prevent spreading of the electron beam in the resist. Low energy electron optical systems are also hard to design for high resolution. Coulomb inter-electron repulsion always becomes more severe for lower electron energy.

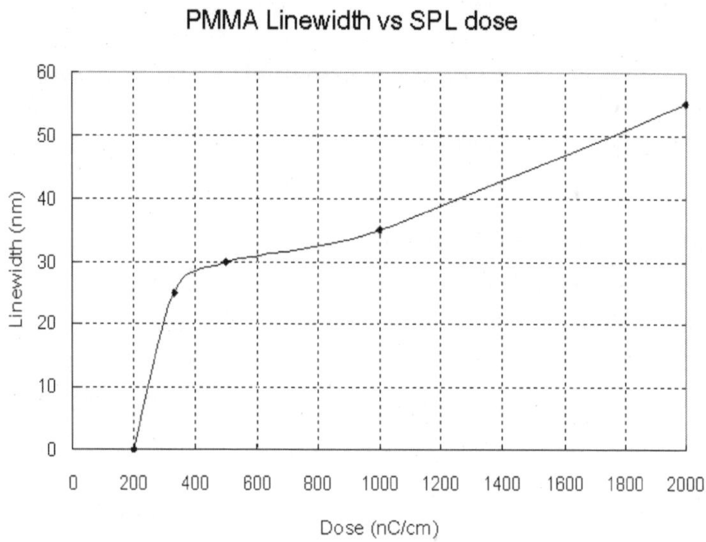

Figure: *Scanning probe lithography. A scanning probe can be used for low-energy electron beam lithography, offering sub-100 nm resolution, determined by the dose of low-energy electrons.*

Another alternative in electron-beam lithography is to use extremely high electron energies (at least 100 keV) to essentially "drill" or sputter the material. This phenomenon has been observed frequently in transmission electron microscopy. However, this is a very inefficient process, due to the inefficient transfer of momentum from the electron beam to the material. As a result it is a slow process, requiring much longer exposure times than conventional electron beam lithography. Also high energy beams always bring up the concern of substrate damage.

Interference lithography using electron beams is another possible path for patterning arrays with nanometre-scale periods. A key advantage of using electrons over photons in interferometry is the much shorter wavelength for the same energy.

Despite the various intricacies and subtleties of electron beam lithography at different energies, it remains the most practical way to concentrate the most energy into the smallest area.

There has been significant interest in the development of multiple electron beam approaches to lithography in order to increase throughput. This work has been supported by SEMATECH and start-up companies such as Multibeam Corporation, Mapper and IMS. However, the degree of parallelism required to be competitive would need to be very high (at least 10 million, as estimated above); this is far in excess of most scheduled demonstrations. A key difficulty is that the total supplied beam current needs to be multiplied by the number of parallel beams(e.g., 10 million), which dramatically increases cost of ownership. Also, the field size does not change, which means increasing the number of beams increases the strength of Coulomb interaction between beams.

Ion Implantation

Ion implantation is a materials engineering process by which ions of a material are accelerated in an electrical field and impacted into a solid. This process is used to change the physical, chemical, or electrical properties of the solid. Ion implantation is used in semiconductor device fabrication and in metal finishing, as well as various applications in materials science research. The ions alter the elemental composition of the target, if the ions differ in composition from the target, stop in the target and stay there.

They also cause many chemical and physical changes in the target by transferring their energy and momentum to the electrons and atomic nuclei of the target material. This causes a structural change, in that the crystal structure of the target can be damaged or even destroyed by the energetic collision cascades. Because the ions have masses comparable to those of the target atoms, they knock the target atoms out of place more than electron beams do. If the ion energy is sufficiently high (usually tens of MeV) to overcome the coulomb barrier, there can even be a small amount of nuclear transmutation.

General Principle

Ion implantation equipment typically consists of an ion source, where ions of the desired element are produced, an accelerator, where the ions are electrostatically accelerated to a high energy, and a target chamber, where the ions impinge on a target, which is the material to be implanted. Thus ion implantation is a special case of particle radiation. Each ion is typically a single atom or molecule, and thus the actual amount of material implanted in the target is the integral over time of the ion current. This amount is called the dose. The currents supplied by implanters are typically small (microamperes), and thus the dose which can be implanted in a reasonable amount of time is small. Therefore, ion implantation finds application in cases where the amount of chemical change required is small.

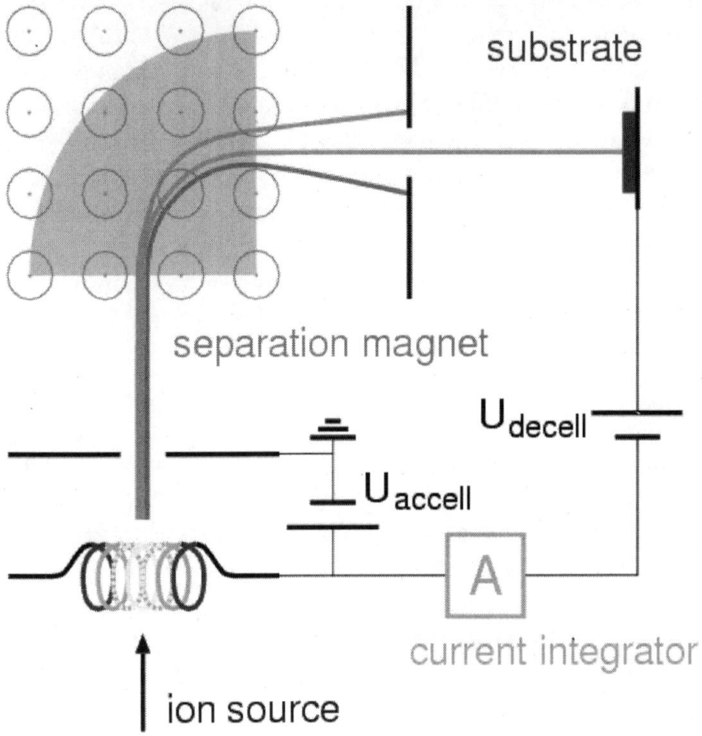

Figure: *Ion implantation setup with mass separator*

Typical ion energies are in the range of 10 to 500 keV (1,600 to 80,000 aJ). Energies in the range 1 to 10 keV (160 to 1,600 aJ) can be used, but result in a penetration of only a few nanometres or less. Energies lower than this result in very little damage to the target, and fall under the designation ion beam deposition. Higher energies can also be used: accelerators capable of 5 MeV (800,000 aJ) are common. However, there is often great structural damage to the target, and because the depth distribution is broad (Bragg peak), the net composition change at any point in the target will be small.

The energy of the ions, as well as the ion species and the composition of the target determine the depth of penetration of the ions in the solid: A monoenergetic ion beam will generally have a broad depth distribution. The average penetration depth is called the range of the ions. Under typical circumstances ion ranges will be between 10 nanometres and 1 micrometre.

Thus, ion implantation is especially useful in cases where the chemical or structural change is desired to be near the surface of the target. Ions gradually lose their energy as they travel through the solid, both from occasional collisions with target atoms (which cause abrupt energy transfers) and from a mild drag from overlap of electron orbitals, which is a continuous process. The loss of ion energy in the target is called stopping and can be simulated with the binary collision approximation method.

Accelerator systems for ion implantation are generally classified into

- Medium current — ion beam currents between 10 μA and ~2 mA.
- High current — ion beam currents up to ~30 mA.
- High energy — ion energies above 200 keV and up to 10 MeV.
- Very high dose — efficient implant of dose greater than 10^{16} ions/cm^2.

All varieties of ion implantation beamline designs contain certain general groups of functional components. The first major segment of an ion beamline includes a device known as an ion source to generate the ion species. The source is closely coupled to biased electrodes for extraction of the ions into the beamline and most often to some means of selecting a particular ion species for transport into the main accelerator section.

The "mass" selection is often accompanied by passage of the extracted ion beam through a magnetic field region with an exit path restricted by blocking apertures, or "slits", that allow only ions with a specific value of the product of mass and velocity/charge to continue down the beamline. If the target surface is larger than the ion beam diameter and a uniform distribution of implanted dose is desired over the target surface, then some combination of beam scanning and wafer motion is used. Finally, the implanted surface is coupled with some method for collecting the accumulated charge of the implanted ions so that the delivered dose can be measured in a continuous fashion and the implant process stopped at the desired dose level.

Application in Semiconductor Device Fabrication

Doping: The introduction of dopants in a semiconductor is the most common application of ion implantation. Dopant ions such as boron, phosphorus or arsenic are generally created from a gas source, so that the purity of the source can be very high. These gases tend to be very hazardous. When implanted in a semiconductor, each dopant atom can create a charge carrier in the semiconductor after annealing. A hole can be created for a p-type dopant, and an electron for an n-type dopant. This modifies the conductivity of the semiconductor in its vicinity. The technique is used, for example, for adjusting the threshold of a MOSFET.

Ion implantation was developed as a method of producing the p-n junction of photovoltaic devices in the late 1970s and early 1980s, along with the use of pulsed-electron beam for rapid annealing, although it has not to date been used for commercial production.

Silicon on Insulator

One prominent method for preparing silicon on insulator (SOI) substrates from conventional silicon substrates is the *SIMOX* (Separation by IMplantation of OXygen) process, wherein a buried high dose oxygen implant is converted to silicon oxide by a high temperature annealing process.

Mesotaxy

Mesotaxy is the term for the growth of a crystallographically matching phase underneath the surface of the host crystal (compare to epitaxy, which is the growth of the matching phase on the surface of a substrate). In this process, ions are implanted at a high enough energy and dose into a material to create a layer of a second phase, and the temperature is controlled so that the crystal structure of the target is not destroyed.

The crystal orientation of the layer can be engineered to match that of the target, even though the exact crystal structure and lattice constant may be very different. For example, after the implantation of nickel ions into a silicon wafer, a layer of nickel silicide can be grown in which the crystal orientation of the silicide matches that of the silicon.

Application in Metal Finishing

Tool Steel Toughening

Nitrogen or other ions can be implanted into a tool steel target (drill bits, for example). The structural change caused by the implantation produces a surface compression in the steel, which prevents crack propagation and thus makes the material more resistant to fracture. The chemical change can also make the tool more resistant to corrosion.

Surface Finishing

In some applications, for example prosthetic devices such as artificial joints, it is desired to have surfaces very resistant to both chemical corrosion and wear due to friction.

Ion implantation is used in such cases to engineer the surfaces of such devices for more reliable performance. As in the case of tool steels, the surface modification caused by ion implantation includes both a surface compression which prevents crack propagation and an alloying of the surface to make it more chemically resistant to corrosion.

Other Applications

Ion Beam Mixing: Ion implantation can be used to achieve ion beam mixing, i.e. mixing up atoms of different elements at an interface. This may be useful for achieving graded interfaces or strengthening adhesion between layers of immiscible materials.

Problems with Ion Implantation

Crystallographic Damage: Each individual ion produces many point defects in the target crystal on impact such as vacancies and interstitials. Vacancies are crystal lattice points unoccupied by an atom: in this case the ion collides with a target atom, resulting in transfer of a significant amount of energy to the target atom such that it leaves its crystal site. This target atom then itself becomes a projectile in the solid, and can cause successive collision events. Interstitials result when such atoms (or the original ion itself) come to rest in the solid, but find no vacant space in the lattice to reside. These point defects can migrate and cluster with each other, resulting in dislocation loops and other defects.

Damage Recovery

Because ion implantation causes damage to the crystal structure of the target which is often unwanted, ion implantation processing is often followed by a thermal annealing. This can be referred to as damage recovery.

Amorphization

The amount of crystallographic damage can be enough to completely amorphize the surface of the target: i.e. it can become an amorphous solid (such a solid produced from a melt is called a glass). In some cases, complete amorphization of a target is preferable to a highly defective crystal: An amorphized film can be regrown at a lower temperature than required to anneal a highly damaged crystal.

Sputtering

Some of the collision events result in atoms being ejected (sputtered) from the surface, and thus ion implantation will slowly etch away a surface. The effect is only appreciable for very large doses.

Ion Channeling

If there is a crystallographic structure to the target, and especially in semiconductor substrates where the crystal structure is more open, particular crystallographic directions offer much lower stopping than other directions. The result is that the range of an ion can be much longer if the ion travels exactly along a particular direction, for example the <110> direction in silicon and other diamond cubic materials. This effect is called *ion channelling*, and, like all the channelling effects, is highly nonlinear, with small variations from perfect orientation resulting in extreme differences in implantation depth. For this reason, most implantation is carried out a few degrees off-axis, where tiny alignment errors will have more predictable effects.

Ion channelling can be used directly in Rutherford backscattering and related techniques as an analytical method to determine the amount and depth profile of damage in crystalline thin film materials.

Hazardous Materials Note

In the ion implantation semiconductor fabrication process of wafers, it is important for the workers to minimize their exposure to the toxic materials used in the ion implanter process. Such hazardous elements, solid source and gasses are used, such as arsine and phosphine. For this reason, the semiconductor fabrication facilities are highly automated, and may feature negative pressure gas bottles safe delivery system (SDS). Other elements may include antimony, arsenic, phosphorus, and boron. Residue of these elements show up when the machine is opened to atmosphere, and can also be accumulated and found concentrated in the vacuum pumps hardware. It is important not to expose yourself to these carcinogenic, corrosive, flammable, and toxic elements. Many overlapping safety protocols must be used when handling these deadly compounds. Use safety, and read MSDSs.

High Voltage Safety

High voltage power supplies in ion implantation equipment can pose a risk of electrocution. In addition, high-energy atomic collisions can generate X-rays and, in some cases, other ionizing radiation and radionuclides. Operators and maintenance personnel should learn and follow the safety advice of the manufacturer and/or the institution responsible for the equipment. Prior to entry to high voltage area, terminal components must be grounded using a grounding stick. Next, power supplies should be locked in the off state and tagged to prevent unauthorized energizing.

Other types of particle accelerator, such as radio frequency linear particle accelerators and laser wakefield plasma accelerators have their own hazards.

6

Gas Detector

A gas detector is a device which detects the presence of various gases within an area, usually as part of a safety system. This type of equipment is used to detect a gas leak and interface with a control system so a process can be automatically shut down. A gas detector can also sound an alarm to operators in the area where the leak is occurring, giving them the opportunity to leave the area. This type of device is important because there are many gases that can be harmful to organic life, such as humans or animals.

Gas detectors can be used to detect combustible, flammable and toxic gases, and oxygen depletion. This type of device is used widely in industry and can be found in a variety of locations such as on oil rigs, to monitor manufacture processes and emerging technologies such as photovoltaic. They may also be used in firefighting.

Types

Gas detectors can be classified according to the operation mechanism (semiconductors, oxidation, catalytic, infrared, etc.). Gas detectors come in two main types: portable devices and fixed gas detectors.

Portable detectors are used to monitor the atmosphere around personnel and are worn on clothing or on a belt/harness. These gas detectors are usually battery operated. They transmit warnings via a series of audible and visible signals such as alarms and flashing lights, when dangerous levels of gas vapors are detected. As detectors measure a gas concentration, the sensor responds to a calibration gas, which serves as the reference point or scale. As a sensor's detection exceeds a preset alarm level, the alarm or signal will be activated. As units, gas detectors are produced as portable or stationary devices. Originally, detectors were produced to detect a single gas, but modern units may detect several toxic or combustible gases, or even a combination of both types.

Fixed type gas detectors may be used for detection of one or more gas types. Fixed type detectors are generally mounted near the process area of a plant or control room. Generally, they are installed on fixed type mild steel structures, and a cable connects the detectors to a SCADA system for continuous monitoring. A tripping interlock can be activated for an emergency situation.

Newer gas analyzers can break up the component signals in a complicated aroma to identify several gasses simultaneously.

Calibration

All gas detectors must be calibrated on a schedule. Of the two types of gas detectors, portables must be calibrated more frequently due to the regular changes in environment they experience. A typical calibration schedule for a fixed system may be quarterly, bi-annually or even annually with some of the more robust units. A typical calibration schedule for a portable gas detector is a daily bump test accompanied by a monthly calibration. Almost every portable gas detector out there has a specific calibration gas requirement which is available from the manufacturer you purchased your monitor from.

Challenge (Bump) Test

As a gas detector is used for employee/worker safety, it is very important to make sure it is operating to manufacturer's specifications. Australian standards specify that a person operating any gas detector is strongly advised to check the gas detector's performance each day, and that it is maintained and used in accordance with the manufacturers instructions and warnings.

A challenge test should consist of exposing the gas detector to a known concentration of gas to ensure that the gas detector will respond and that both the audible and visual alarms activate. It is also important inspect the gas detector for any accidental or deliberate damage by checking that the housing and screws are all intact to prevent any liquid ingress and that the filter is clean, all of which can affect the functionality of the gas detector. The basic calibration or challenge test kit will consist of: Calibration Gas / Regulator / Calibration Cap and hose (generally supplied with the gas detector at time of purchase) and a case for storage and transport. Because 1 in every 2,500 untested instruments will fail to respond to a dangerous concentration of gas, many large businesses will utilise an automated test/calibration station for use to bump test and calibrate their gas detectors daily.

Oxygen Concentration

Oxygen deficiency gas monitors are used for employee and workforce safety. Cryogenic substances such as liquid nitrogen (LN2), liquid helium (He), and liquid argon (Ar) are inert and can displace oxygen (O_2) in a confined space if a leak is present. A rapid decrease of oxygen can provide a very dangerous environment for employees, who may not notice this problem before they suddenly lose consciousness. With this in mind, an oxygen gas monitor is important to have when cryogenics are present. Laboratories, MRI rooms, pharmaceutical, semiconductor, and cryogenic suppliers are typical users of oxygen monitors.

Oxygen fraction in a breathing gas is measured by electro-galvanic fuel cell sensors. They may be used stand-alone, for example to determine the proportion of oxygen in a nitrox mixture used in scuba diving, or as part of feedback loop which maintains a constant partial pressure of oxygen in a rebreather.

Hydrocarbons and VOCs

Detection of hydrocarbons can be based on the mixing properties of gaseous hydrocarbons – or other volatile organic compounds (VOCs) – and the sensing material incorporated in the sensor. The selectivity and sensitivity depends on the molecular structure of the VOC and the concentration; however it is difficult to design a selective sensor for a single VOC. Many VOC sensors detect using a fuel-cell technique.

VOCs in the environment or certain atmospheres can be detected based on different principles and interactions between the organic compounds and the sensor components. There are electronic devices that can detect ppm concentrations despite not being particularly selective. Others can predict with reasonable accuracy the molecular structure of the volatile organic compounds in the environment or enclosed atmospheres and could be used as accurate monitors of the Chemical Fingerprint and further as health monitoring devices.

Solid-phase microextraction(SPME) techniques are used to collect VOCs at low concentrations for analysis.

Considerations for Detection of Hydrocarbon Gases /Risk Control

- Methane is lighter than air (possibility of accumulation under roofs)
- Ethane is slightly heavier than air (possibility of pooling at ground levels / pits)
- Propane is heavier than air (possibility of pooling at ground levels / pits)
- Butane is heavier than air (possibility of pooling at ground levels / pits)

Combustible

- Catalytic bead sensor
- Explosimetre
- Infrared point sensor
- Infrared open path detector

Other

- Flame ionization detector
- Nondispersive infrared sensor
- Photoionization detector
- Zirconium oxide sensor cell
- Catalytic sensors
- Metal oxide semiconductor
- Gold film
- Colourimetric Detector Tubes
- Sample collection and chemical analysis
- Piezoelectric microcantilever
- Holographic Sensor
- Thermal Conductivity Detector
- Electrochemical gas sensor

Research

The European Community has supported research called the MINIGAS project that was coordinated by VTT Technical Research Centre of Finland. This research project aims to develop new types of photonics-based gas sensors, and to support the creation of smaller instruments with equal or higher speed and sensitivity than conventional laboratory-grade gas detectors.

Gas Leak Detection

Gas leak detection is the process of identifying potentially hazardous gas leaks by means of various sensors. These sensors usually employ an audible alarm to alert people when a dangerous gas has been detected. Common sensors used today include infrared point sensors, ultrasonic sensors, electrochemical gas sensors, and semiconductor sensors. More recently, infrared imaging sensors have come into use. All of these sensors are used for a wide range of applications, and can be found in industrial plants, refineries, wastewater treatment facilities, vehicles, and around the home.

Gas leak detection methods became a concern after the effects of harmful gases on human health were discovered. Before modern electronic sensors, early detection methods relied on less precise detectors. Through the 19th and early 20th centuries, coal miners would bring canaries down to the tunnels with them as an early detection system against life threatening gases such as carbon dioxide, carbon monoxide and methane. The canary, normally a very songful bird, would stop singing and eventually die in the presence of these gases, signaling the miners to exit the mine quickly.

Before the development of electronic household carbon monoxide detectors in the 1980s and 90s, carbon monoxide presence was detected with a chemically infused paper that turned brown when exposed to the gas. Since then, many technologies and devices have been developed to detect, monitor, and alert the leakage of a wide array of gases.

Types of Gas Detectors

Electrochemical: Electrochemical gas detectors work by allowing gases to diffuse through a porous membrane to an electrode where it is either oxidized or reduced. The amount of current produced is determined by how much of the gas is oxidized at the electrode. The sensor is then able to determine the concentration of the gas. Manufactures can customize electrochemical gas detectors by changing the porous barrier to allow for the detection of a certain gas concentration range. Also, since the diffusion barrier is a physical/mechanical barrier, the detector tends to be more stable and reliable over the sensor's duration and thus requires less maintenance than other types of detectors.

However, the sensors themselves are subject to corrosive elements or chemical contamination, and may last only 1–2 years before a replacement is required. Electrochemical gas detectors are used in a wide variety of environments such as refineries, gas turbines, chemical plants, underground gas storage facilities, and more.

Infrared Point

Infrared (IR) point sensors use radiation passing through a volume of gas to detect leaks. Energy from the radiation is absorbed as it passes through the gas at certain wavelengths. The range of wavelengths that is absorbed depends on the properties of the specific gas. Carbon monoxide absorbs wavelengths of about 4.2-4.5 μm, for example. This is approximately a factor of 10 larger than the wavelength of visible light, which ranges from .39 μm to .75 μm for most people. The energy in this wavelength is compared to a wavelength outside of the absorption range; the difference in energy between these two wavelengths is proportional to the concentration of gas present.

This type of sensor is advantageous because it does not have to be placed in the gas itself in order to detect it. Infrared point sensors can be used to detect hydrocarbons, compounds composed of hydrogen and carbon atoms, and other infrared active gases such as water vapor and calcium fluoride.

IR sensors are commonly found in wastewater treatment facilities, refineries, gas turbines, chemical plants, and other facilities where flammable gases are present and the possibility of an explosion exists. Engine emissions are another area where IR sensors are being researched for use. The sensor would be able to detect high levels of carbon dioxide in the vehicles' exhaust, and even be integrated with the vehicles' electronic systems to notify drivers.

Infrared Imaging

Infrared point sensors are active measurement systems, in that they typically measure the absorption of a gas by passing it through a laser-illuminated chamber and measuring the change in transmitted signal. Infrared imaging sensors include both active and passive systems.

For active sensing, IR imaging sensors typically scan a laser across the field of view of a scene and look for backscattered light at the absorption line wavelength of a specific target gas. Passive IR imaging sensors, on the other hand, measure spectral changes at each pixel in an image and look for specific spectral signatures which indicate the presence of target gases. The types of compounds which can be imaged are the same as those which can be detected with infrared point detectors.

Semiconductor

Semiconductor sensors detect gases by a chemical reaction that takes place when the gas comes in contact with the sensor. Tin dioxide is the most common material used in semiconductor sensors, and the electrical resistance in the sensor is decreased when it comes in contact with the monitored gas. The resistance of the tin dioxide is typically around 50 kΩ in air but can drop to around 3.5 kΩ in the presence of 1% methane. This change in resistance is used to calculate the gas concentration. Semiconductor sensors are commonly used to detect hydrogen, oxygen, alcohol, and harmful gases such as carbon monoxide. One of the most common uses for semiconductor sensors is in carbon monoxide sensors. They are also used in breathalyzers. Because the sensor must come in contact with the gas in order to detect it, semiconductor sensors work over a smaller distance than infrared point or ultrasonic detectors.

Ultrasonic

Ultrasonic gas detectors use acoustic sensors to detect changes in the background noise of its environment. Since most gas leaks occur in the ultrasonic range of 25 kHz to 10 MHz, the sensors are able to easily distinguish these frequencies from background noise which occurs in the audible range of 20 Hz to 20 kHz. The ultrasonic gas leak detector then produces an alarm when there is an ultrasonic deviation from the normal condition of background noise. Despite the fact that Ultrasonic gas leak detectors don't measure gas concentration, the device is still able to determine the leak rate of an escaping gas. By measuring its ultrasonic sound level, the detector is able to determine the leak rate, which depends on the gas pressure and size of the leak.

The bigger the leak, the larger its ultrasonic sound level will be. Ultrasonic gas detectors are mainly used for outdoor environments where weather conditions can easily dissipate escaping gas before allowing it to reach gas leak detectors that require contact with the gas in order to detect it and sound an alarm. These detectors are commonly found on offshore and onshore oil/gas platforms, gas compressor and metring stations, gas turbine power plants, and other facilities that house a lot of outdoor pipeline.

Holographic

Holographic gas sensors use light reflection to detect changes in a polymer film matrix containing a hologram. Since holograms reflect light at certain wavelengths a change in their composition can generate a colourful reflection indicative of the presence of a gas molecule. However, holographic sensors require illumination sources such as white light or lasers, and an observer or CCD detector.

Household Safety

There are many different sensors that can be purchased to detect hazardous gases around the house. Carbon monoxide is a very dangerous gas that robs the lungs of oxygen, killing hundreds of people worldwide each year. It is an odorless, colourless gas, making it impossible for humans to detect it. Carbon monoxide detectors can be purchased for around US$20–60. Handheld flammable gas detectors can be used to trace leaks from natural gas lines, propane tanks, butane tanks, or any other combustible gas.

Holographic Sensor

A holographic sensor is a device that comprises a hologram embedded in a smart material that detects certain molecules or metabolites. This detection is usually a chemical interaction that is transduced as a change in one of the properties of the holographic reflection (as in the Bragg reflector), either refractive index or spacing between the holographic fringes. The specificity of the sensor can be controlled by adding molecules in the polymer film that selectively interacts with the molecules of interest.

A holographic sensor aims to integrate the sensor component, the transducer and the display in one device for fast reading of molecular concentrations based in colourful reflections or wavelengths.

Certain molecules that mimic biomolecule active sites or binding sites can be incorporated into the polymer that forms the holographic film in order to make the holographic sensors selective and/or sensitive to certain medical important molecules like glucose, etc.

The holographic sensors can be read from a fair distance because the transducer element is light that has been refracted and reflected by the holographic grating embedded in the sensor. Therefore they can be used in industrial applications where non-contact with the sensor is required. Other applications for holographic sensors are anti counterfeiting

Metabolites

Some of the metabolites detected by a holographic sensor are:

- pH
- Hydrocarbons
- VOCs
- Gases
- Glucose
- Water content
- Lactate and other biomolecules
- Metal ions

Antibodies in Sensing Applications

An antibody (Ab), also known as an immunoglobulin (Ig), is a large Y-shape protein produced by plasma cells that is used by the immune system to identify and neutralize foreign objects such as bacteria and viruses. The antibody recognises a unique part of the foreign target, called an antigen. Each tip of the "Y" of an antibody contains a paratope (a structure analogous to a lock) that is specific for one particular epitope (similarly analogous to a key) on an antigen, allowing these two structures to bind together with precision. Using this binding mechanism, an antibody can *tag* a microbe or an infected cell for attack by other parts of the immune system, or can neutralize its target directly (for example, by blocking a part of a microbe that is essential for its invasion and survival). The production of antibodies is the main function of the humoral immune system.

Antibodies are secreted by a type of white blood cell called a plasma cell. Antibodies can occur in two physical forms, a soluble form that is secreted from the cell, and a membrane-bound form that is attached to the surface of a B cell and is referred to as the B cell receptor (BCR). The BCR is only found on the surface of B cells and facilitates the activation of these cells and their subsequent differentiation into either antibody factories called plasma cells or memory B cells that will survive in the body and remember that same antigen so the B cells can respond faster upon future exposure. In most cases, interaction of the B cell with a T helper cell is necessary to produce full activation of the B cell and, therefore, antibody generation following antigen binding. Soluble antibodies are released into the blood and tissue fluids, as well as many secretions to continue to survey for invading microorganisms.

Antibodies are glycoproteins belonging to the immunoglobulin superfamily; the terms *antibody* and *immunoglobulin* are often used interchangeably. Antibodies are typically made of basic structural units—each with two large heavy chains and two small light chains. There are several different types of antibody heavy chains, and several different kinds of antibodies, which are grouped into different *isotypes* based on which heavy chain they possess. Five different antibody isotypes are known in mammals, which perform different roles, and help direct the appropriate immune response for each different type of foreign object they encounter.

Though the general structure of all antibodies is very similar, a small region at the tip of the protein is extremely variable, allowing millions of antibodies with slightly different tip structures, or antigen-binding sites, to exist. This region is known as the *hypervariable region*. Each of these variants can bind to a different antigen. This enormous diversity of antibodies allows the immune system to recognise an equally wide variety of antigens. The large and diverse population of antibodies is generated by random combinations of a set of gene segments that encode different antigen-binding sites (or *paratopes*), followed by random mutations in this area of the antibody gene, which create further diversity. Antibody genes also re-organise in a process called class switching that changes the base of the heavy chain to another, creating a different isotype of the antibody that retains the antigen-specific variable region. This allows a single antibody to be used by several different parts of the immune system.

Forms

The membrane-bound form of an antibody may be called a surface immunoglobulin (sIg) or a membrane immunoglobulin (mIg). It is part of the B cell receptor (BCR), which allows a B cell to detect when a specific antigen is present in the body and triggers B cell activation. The BCR is composed of surface-bound IgD or IgM antibodies and associated Ig-α and Ig-β heterodimers, which are capable of signal transduction. A typical human B cell will have 50,000 to 100,000 antibodies bound to its surface. Upon antigen binding, they cluster in large patches, which can exceed 1 micrometre in diameter, on lipid rafts that isolate the BCRs from most other cell signaling receptors. These patches may improve the efficiency of the cellular immune response. In humans, the cell surface is bare around the B cell receptors for several hundred nanometres, which further isolates the BCRs from competing influences.

Isotypes

Antibodies can come in different varieties known as isotypes or classes. In placental mammals there are five antibody isotypes known as IgA, IgD, IgE, IgG, and IgM. They are each named with an "Ig" prefix that stands for immunoglobulin, another name for antibody, and differ in their biological properties, functional locations and ability to deal with different antigens, as depicted in the table.

Antibody isotypes of mammals		
Name	***Types***	***Description***
IgA	2	Found in mucosal areas, such as the gut, respiratory tract and urogenital tract, and prevents colonization by pathogens. Also found in saliva, tears, and breast milk.
IgD	1	Functions mainly as an antigen receptor on B cells that have not been exposed to antigens. It has been shown to activate basophils and mast cells to produce antimicrobial factors.
IgE	1	Binds to allergens and triggers histamine release from mast cells and basophils, and is involved in allergy. Also protects against parasitic worms.
IgG	4	In its four forms, provides the majority of antibody-based immunity against invading pathogens. The only antibody capable of crossing the placenta to give passive immunity to the fetus.
IgM	1	Expressed on the surface of B cells (monomer) and in a secreted form (pentamer) with very high avidity. Eliminates pathogens in the early stages of B cell-mediated (humoral) immunity before there is sufficient IgG.

The antibody isotype of a B cell changes during cell development and activation. Immature B cells, which have never been exposed to an antigen, express only the IgM+ isotype in a cell surface bound form. The B lymphocyte, in its mature ready-to-respond form, is known as "naive B lymphocyte." The naive B lymphocyte express both surface IgM+ and IgD+. The co-expression of both these immunoglobulin isotypes renders the B cell 'mature' and ready to respond to antigen. B cell activation follows engagement of the cell-bound antibody molecule with an antigen, causing the cell to divide and differentiate into an antibody-producing cell called a plasma cell. In this activated form, the B cell starts to produce antibody in a secreted form rather than a membrane-bound form. Some daughter cells of the activated B cells undergo isotype switching, a mechanism that causes the production of antibodies to change from IgM or IgD to the other antibody isotypes, IgE, IgA, or IgG, that have defined roles in the immune system.

Structure

Antibodies are heavy (~150 kDa) globular plasma proteins. They have sugar chains added to some of their amino acid residues. In other words, antibodies are glycoproteins. The basic functional unit of each antibody is an immunoglobulin (Ig) monomer (containing only one Ig unit); secreted antibodies can also be dimeric with two Ig units as with IgA, tetrameric with four Ig units like teleost fish IgM, or pentameric with five Ig units, like mammalian IgM.

Figure: *Several immunoglobulin domains make up the two heavy chains (red and blue) and the two light chains (green and yellow) of an antibody. The immunoglobulin domains are composed of between 7 (for constant domains) and 9 (for variable domains) β-strands.*

The variable parts of an antibody are its V regions, and the constant part is its C region.

Immunoglobulin Domains

The Ig monomer is a "Y"-shaped molecule that consists of four polypeptide chains; two identical *heavy chains* and two identical *light chains* connected by disulfide bonds. Each chain is composed of structural domains called immunoglobulin domains. These domains contain about 70–110 amino acids and are classified into different categories (for example, variable or IgV, and constant or IgC) according to their size and function. They have a characteristic immunoglobulin fold in which two beta sheets create a "sandwich" shape, held together by interactions between conserved cysteines and other charged amino acids.

Heavy Chain

There are five types of mammalian Ig heavy chain denoted by the Greek letters: α, δ, ε, γ, and μ. The type of heavy chain present defines the *class* of antibody; these chains are found in IgA, IgD, IgE, IgG, and IgM antibodies, respectively. Distinct heavy chains differ in size and composition; α and γ contain approximately 450 amino acids, whereas μ and ε have approximately 550 amino acids.

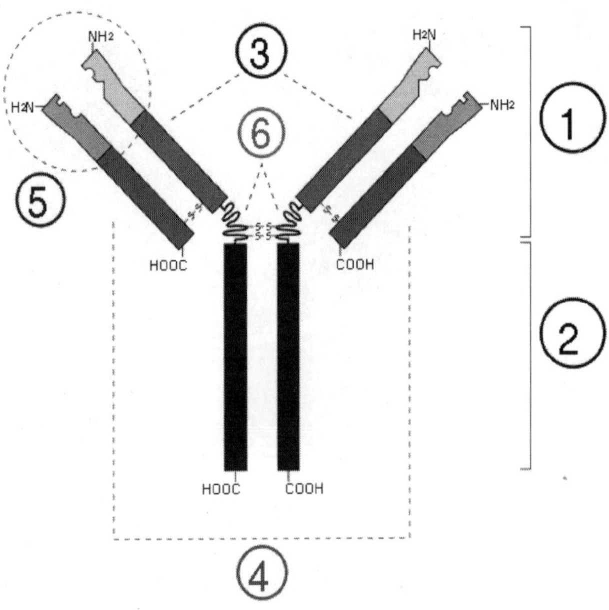

Figure: *1. Fab region; 2. Fc region; 3. Heavy chain (blue) with one variable (V_H) domain followed by a constant domain (C_H1), a hinge region, and two more constant (C_H2 and C_H3) domains; 4. Light chain (green) with one variable (V_L) and one constant (C_L) domain; 5. Antigen binding site (paratope); 6. Hinge regions.*

In birds, the major serum antibody, also found in yolk, is called IgY. It is quite different from mammalian IgG. However, in some older literature and even on some commercial life sciences product websites it is still called "IgG", which is incorrect and can be confusing.

Each heavy chain has two regions, the *constant region* and the *variable region*. The constant region is identical in all antibodies of the same isotype, but differs in antibodies of different isotypes. Heavy chains γ, α and δ have a constant region composed of *three* tandem (in a line) Ig domains, and a hinge region for added flexibility; heavy chains μ and ε have a constant region composed of *four* immunoglobulin domains. The variable region of the heavy chain differs in antibodies produced by different B cells, but is the same for all antibodies produced by a single B cell or B cell clone. The variable region of each heavy chain is approximately 110 amino acids long and is composed of a single Ig domain.

Light Chain

In mammals there are two types of immunoglobulin light chain, which are called lambda (λ) and kappa (κ). A light chain has two successive domains: one constant domain and one variable domain. The approximate length of a light chain is 211 to 217 amino acids. Each antibody contains two light chains that are always identical; only one type of light chain, κ or λ, is present per antibody in mammals. Other types of light chains, such as the iota (θ) chain, are found in other vertebrates like sharks (Chondrichthyes) and bony fishes (Teleostei).

CDRs, Fv, Fab and Fc Regions

Some parts of an antibody have the same functions. The arms of the Y, for example, contain the sites that can bind two antigens (in general, identical) and, therefore, recognise specific foreign objects. This region of the antibody is called the *Fab (fragment, antigen-binding) region*. It is composed of one constant and one variable domain from each heavy and light chain of the antibody. The paratope is shaped at the amino terminal end of the antibody monomer by the variable domains from the heavy and light chains. The variable domain is also referred to as the F_V region and is the most important region for binding to antigens. To be specific, variable loops of β-strands, three each on the light (V_L) and heavy (V_H) chains are responsible for binding to the antigen. These loops are referred to as the complementarity determining regions (CDRs). The structures of these CDRs have been clustered and classified by Chothia et al. and more recently by North et al. In the framework of the immune network theory, CDRs are also called idiotypes. According to immune network theory, the adaptive immune system is regulated by interactions between idiotypes.

The base of the Y plays a role in modulating immune cell activity. This region is called the *Fc (Fragment, crystallizable) region*, and is composed of two heavy chains that contribute two or three constant domains depending on the class of the antibody. Thus, the Fc region ensures that each antibody generates an appropriate immune response for a given antigen, by binding to a specific class of Fc receptors, and other immune molecules, such as complement proteins. By doing this, it mediates different physiological effects including recognition of opsonized particles, lysis of cells, and degranulation of mast cells, basophils, and eosinophils.

Function

Activated B cells differentiate into either antibody-producing cells called plasma cells that secrete soluble antibody or memory cells that survive in the body for years afterward in order to allow the immune system to remember an antigen and respond faster upon future exposures.

At the prenatal and neonatal stages of life, the presence of antibodies is provided by passive immunization from the mother. Early endogenous antibody production varies for different kinds of antibodies, and usually appear within the first years of life. Since antibodies exist freely in the bloodstream, they are said to be part of the humoral immune system. Circulating antibodies are produced by clonal B cells that specifically respond to only one antigen (an example is a virus capsid protein fragment). Antibodies contribute to immunity in three ways: They prevent pathogens from entering or damaging cells by binding to them; they stimulate removal of pathogens by macrophages and other cells by coating the pathogen; and they trigger destruction of pathogens by stimulating other immune responses such as the complement pathway.

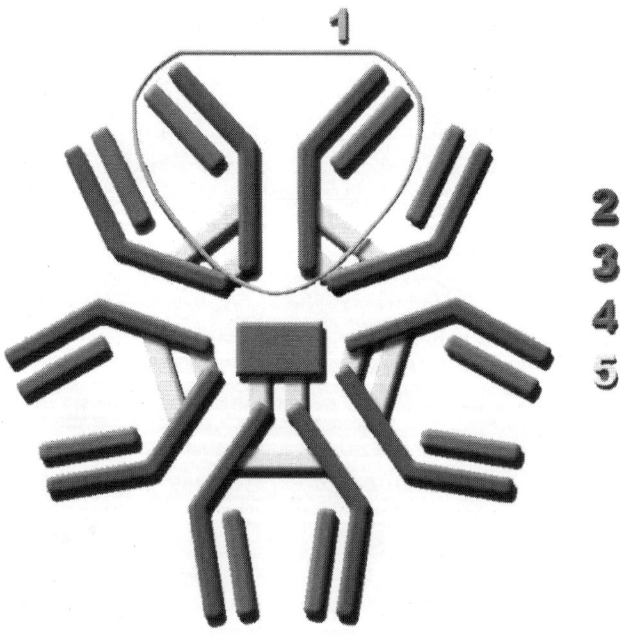

Figure: *The secreted mammalian IgM has five Ig units. Each Ig unit (labeled 1) has two epitope binding Fab regions, so IgM is capable of binding up to 10 epitopes.*

Activation of Complement

Antibodies that bind to surface antigens (like on bacteria) will attract the first component of the complement cascade with their Fc region and initiate activation of the "classical" complement system. This results in the killing of bacteria in two ways. First, the binding of the antibody and complement molecules marks the microbe for ingestion by phagocytes in a process called opsonization; these phagocytes are attracted by certain complement molecules generated in the complement cascade. Second, some complement system components form a membrane attack complex to assist antibodies to kill the bacterium directly. An antibody is a molecule that specificially inactivates an antigen.

Activation of Effector Cells

To combat pathogens that replicate outside cells, antibodies bind to pathogens to link them together, causing them to agglutinate. Since an antibody has at least two paratopes, it can bind more than one antigen by binding identical epitopes carried on the surfaces of these antigens. By coating the pathogen, antibodies stimulate effector functions against the pathogen in cells that recognise their Fc region.

Those cells that recognise coated pathogens have Fc receptors, which, as the name suggests, interacts with the Fc region of IgA, IgG, and IgE antibodies. The engagement of a particular antibody with the Fc receptor on a particular cell triggers an effector function of that cell; phagocytes will phagocytose, mast cells and neutrophils will degranulate, natural killer cells will release cytokines and cytotoxic molecules; that will ultimately result in destruction of the invading microbe. The Fc receptors are isotype-specific, which gives greater flexibility to the immune system, invoking only the appropriate immune mechanisms for distinct pathogens.

Natural Antibodies

Humans and higher primates also produce "natural antibodies" that are present in serum before viral infection. Natural antibodies have been defined as antibodies that are produced without any previous infection, vaccination, other foreign antigen exposure or passive immunization. These antibodies can activate the classical complement pathway leading to lysis of enveloped virus particles long before the adaptive immune response is activated. Many natural antibodies are directed against the disaccharide galactose α(1,3)-galactose (α-Gal), which is found as a terminal sugar on glycosylated cell surface proteins, and generated in response to production of this sugar by bacteria contained in the human gut. Rejection of xenotransplantated organs is thought to be, in part, the result of natural antibodies circulating in the serum of the recipient binding to α-Gal antigens expressed on the donor tissue.

Immunoglobulin Diversity

Virtually all microbes can trigger an antibody response. Successful recognition and eradication of many different types of microbes requires diversity among antibodies; their amino acid composition varies allowing them to interact with many different antigens. It has been estimated that humans generate about 10 billion different antibodies, each capable of binding a distinct epitope of an antigen. Although a huge repertoire of different antibodies is generated in a single individual, the number of genes available to make these proteins is limited by the size of the human genome. Several complex genetic mechanisms have evolved that allow vertebrate B cells to generate a diverse pool of antibodies from a relatively small number of antibody genes.

Domain Variability

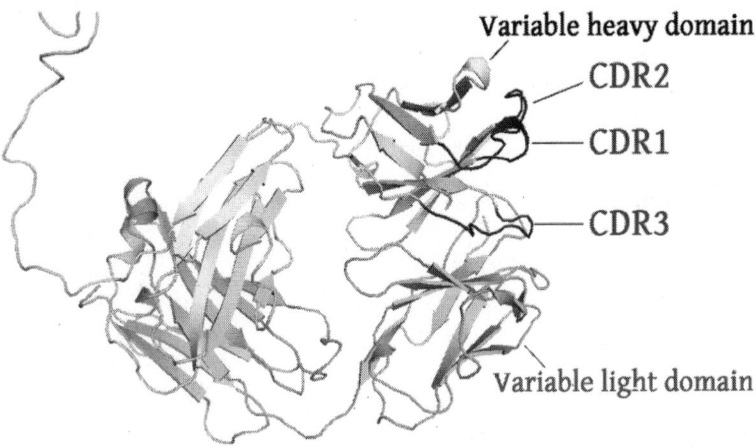

Figure: *The complementarity determining regions of the heavy chain are shown in red (PDB 1IGT)*

The region (locus) of a chromosome that encodes an antibody is large and contains several distinct genes for each domain of the antibody—the locus containing heavy chain genes (IGH@) is found on chromosome 14, and the loci containing lambda and kappa light chain genes (IGL@ and IGK@) are found on chromosomes 22 and 2 in humans. One of these domains is called the variable domain, which is present in each heavy and light chain of every antibody, but can differ in different antibodies generated from distinct B cells. Differences, between the variable domains, are located on three loops known as hypervariable regions (HV-1, HV-2 and HV-3) or complementarity determining regions (CDR1, CDR2 and CDR3). CDRs are supported within the variable domains by conserved framework regions. The heavy chain locus contains about 65 different variable domain genes that all differ in their CDRs. Combining these genes with an array of genes for other domains of the antibody generates a large cavalry of antibodies with a high degree of variability. This combination is called V(D)J recombination discussed below.

V(D)J Recombination

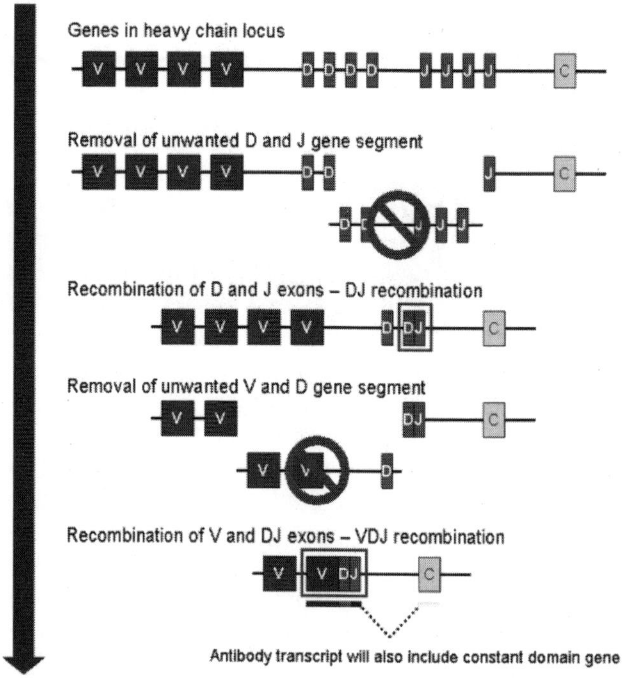

Figure: *Simplified overview of V(D)J recombination of immunoglobulin heavy chains*

Somatic recombination of immunoglobulins, also known as *V(D)J recombination*, involves the generation of a unique immunoglobulin variable region. The variable region of each immunoglobulin heavy or light chain is encoded in several pieces—known as gene segments (subgenes). These segments are called variable (V), diversity (D) and joining (J) segments. V, D and J segments are found in Ig heavy chains, but only V and J segments are found in Ig light chains. Multiple copies of the V, D and J gene segments exist, and are tandemly arranged in the genomes of mammals. In the bone marrow, each developing B cell will assemble an immunoglobulin variable region by randomly selecting and combining one V, one D and one J gene segment (or one V and one J segment in the light chain). As there are multiple copies of each type of gene segment, and different combinations of gene segments can be used to generate each immunoglobulin variable region, this process generates a huge number of antibodies, each with different paratopes, and thus different antigen specificities. Interestingly, the rearrangement of several subgenes (i.e. V2 family) for lambda light chain immunoglobulin is coupled with the activation of microRNA miR-650, which further influences biology of B-cells.

After a B cell produces a functional immunoglobulin gene during V(D)J recombination, it cannot express any other variable region (a process known as allelic exclusion) thus each B cell can produce antibodies containing only one kind of variable chain.

Somatic Hypermutation and Affinity Maturation

Following activation with antigen, B cells begin to proliferate rapidly. In these rapidly dividing cells, the genes encoding the variable domains of the heavy and light chains undergo a high rate of point mutation, by a process called *somatic hypermutation* (SHM). SHM results in approximately one nucleotide change per variable gene, per cell division. As a consequence, any daughter B cells will acquire slight amino acid differences in the variable domains of their antibody chains. This serves to increase the diversity of the antibody pool and impacts the antibody's antigen-binding affinity. Some point mutations will result in the production of antibodies that have a weaker interaction (low affinity) with their antigen than the original antibody, and some mutations will generate antibodies with a stronger interaction (high affinity). B cells that express high affinity antibodies on their surface will receive a strong survival signal during interactions with other cells, whereas those with low affinity antibodies will not, and will die by apoptosis. Thus, B cells expressing antibodies with a higher affinity for the antigen will outcompete those with weaker affinities for function and survival. The process of generating antibodies with increased binding affinities is called *affinity maturation*. Affinity maturation occurs in mature B cells after V(D)J recombination, and is dependent on help from helper T cells.

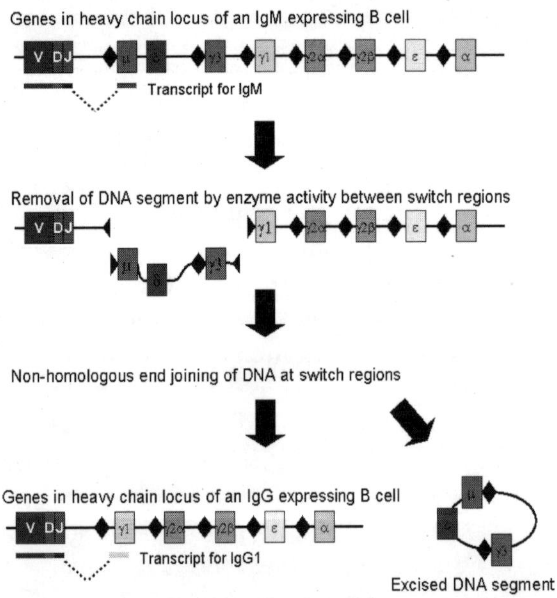

***Figure:** Mechanism of class switch recombination that allows isotype switching in activated B cells*

Class Switching

Isotype or class switching is a biological process occurring after activation of the B cell, which allows the cell to produce different classes of antibody (IgA, IgE, or IgG). The different classes of antibody, and thus effector functions, are defined by the constant (C) regions of the immunoglobulin heavy chain. Initially, naive B cells express only cell-surface IgM and IgD with identical antigen binding regions. Each isotype is adapted for a distinct function; therefore, after activation, an antibody with an IgG, IgA, or IgE effector function might be required to effectively eliminate an antigen. Class switching allows different daughter cells from the same activated B cell to produce antibodies of different isotypes. Only the constant region of the antibody heavy chain changes during class switching; the variable regions, and therefore antigen specificity, remain unchanged. Thus the progeny of a single B cell can produce antibodies, all specific for the same antigen, but with the ability to produce the effector function appropriate for each antigenic challenge. Class switching is triggered by cytokines; the isotype generated depends on which cytokines are present in the B cell environment.

Class switching occurs in the heavy chain gene locus by a mechanism called class switch recombination (CSR). This mechanism relies on conserved nucleotide motifs, called *switch (S) regions*, found in DNA upstream of each constant region gene (except in the δ-chain). The DNA strand is broken by the activity of a series of enzymes at two selected S-regions. The variable domain exon is rejoined through a process called non-homologous end joining (NHEJ) to the desired constant region (γ, α or ε). This process results in an immunoglobulin gene that encodes an antibody of a different isotype.

Affinity Designations

A group of antibodies can be called *monovalent* (or *specific*) if they have affinity for the same epitope, or for the same antigen (but potentially different epitopes on the molecule), or for the same strain of microorganism (but potentially different antigens on or in it). In contrast, a group of antibodies can be called *polyvalent* (or *unspecific*) if they have affinity for various antigens or microorganisms. Intravenous immunoglobulin, if not otherwise noted, consists of polyvalent IgG. In contrast, monoclonal antibodies are monovalent for the same epitope.

Medical Applications

Disease Diagnosis and Therapy

Detection of particular antibodies is a very common form of medical diagnostics, and applications such as serology depend on these methods. For example, in biochemical assays for disease diagnosis, a titer of antibodies directed against Epstein-Barr virus or Lyme disease is estimated from the blood. If those antibodies are not present, either the person is not infected or the infection occurred a *very* long time ago, and the B cells generating these specific antibodies have naturally decayed.

In clinical immunology, levels of individual classes of immunoglobulins are measured by nephelometry (or turbidimetry) to characterize the antibody profile of patient. Elevations in different classes of immunoglobulins are sometimes useful in determining the cause of liver damage in patients for whom the diagnosis is unclear. For example, elevated IgA indicates alcoholic cirrhosis, elevated IgM indicates viral hepatitis and primary biliary cirrhosis, while IgG is elevated in viral hepatitis, autoimmune hepatitis and cirrhosis. Autoimmune disorders can often be traced to antibodies that bind the body's own epitopes; many can be detected through blood tests. Antibodies directed against red blood cell surface antigens in immune mediated hemolytic anemia are detected with the Coombs test. The Coombs test is also used for antibody screening in blood transfusion preparation and also for antibody screening in antenatal women. Practically, several immunodiagnostic methods based on detection of complex antigen-antibody are used to diagnose infectious diseases, for example ELISA, immunofluorescence, Western blot, immunodiffusion, immunoelectrophoresis, and magnetic immunoassay. Antibodies raised against human chorionic gonadotropin are used in over the counter pregnancy tests. Targeted monoclonal antibody therapy is employed to treat diseases such as rheumatoid arthritis, multiple sclerosis, psoriasis, and many forms of cancer including non-Hodgkin's lymphoma, colorectal cancer, head and neck cancer and breast cancer. Some immune deficiencies, such as X-linked agammaglobulinemia and hypogammaglobulinemia, result in partial or complete lack of antibodies. These diseases are often treated by inducing a short term form of immunity called passive immunity. Passive immunity is achieved through the transfer of ready-made antibodies in the form of human or animal serum, pooled immunoglobulin or monoclonal antibodies, into the affected individual.

Prenatal Therapy

Rhesus factor, also known as Rhesus D (RhD) antigen, is an antigen found on red blood cells; individuals that are Rhesus-positive (Rh+) have this antigen on their red blood cells and individuals that are Rhesus-negative (Rh–) do not. During normal childbirth, delivery trauma or complications during pregnancy, blood from a fetus can enter the mother's system. In the case of an Rh-incompatible mother and child, consequential blood mixing may sensitize an Rh- mother to the Rh antigen on the blood cells of the Rh+ child, putting the remainder of the pregnancy, and any subsequent pregnancies, at risk for hemolytic disease of the newborn.

Rho(D) immune globulin antibodies are specific for human Rhesus D (RhD) antigen. Anti-RhD antibodies are administered as part of a prenatal treatment regimen to prevent sensitization that may occur when a Rhesus-negative mother has a Rhesus-positive fetus.

Treatment of a mother with Anti-RhD antibodies prior to and immediately after trauma and delivery destroys Rh antigen in the mother's system from the fetus. It is important to note that this occurs before the antigen can stimulate maternal B cells to "remember" Rh antigen by generating memory B cells. Therefore, her humoral immune system will not make anti-Rh antibodies, and will not attack the Rhesus antigens of the current or subsequent babies. Rho(D) Immune Globulin treatment prevents sensitization that can lead to Rh disease, but does not prevent or treat the underlying disease itself.

Research Applications

Specific antibodies are produced by injecting an antigen into a mammal, such as a mouse, rat, rabbit, goat, sheep, or horse for large quantities of antibody. Blood isolated from these animals contains *polyclonal antibodies*—multiple antibodies that bind to the same antigen—in the serum, which can now be called antiserum.

Antigens are also injected into chickens for generation of polyclonal antibodies in egg yolk. To obtain antibody that is specific for a single epitope of an antigen, antibody-secreting lymphocytes are isolated from the animal and immortalized by fusing them with a cancer cell line. The fused cells are called hybridomas, and will continually grow and secrete antibody in culture. Single hybridoma cells are isolated by dilution cloning to generate cell clones that all produce the same antibody; these antibodies are called *monoclonal antibodies*. Polyclonal and monoclonal antibodies are often purified using Protein A/G or antigen-affinity chromatography.

In research, purified antibodies are used in many applications. They are most commonly used to identify and locate intracellular and extracellular proteins. Antibodies are used in flow cytometry to differentiate cell types by the proteins they express; different types of cell express different combinations of cluster of differentiation molecules on their surface, and produce different intracellular and secretable proteins. They are also used in immunoprecipitation to separate proteins and anything bound to them (co-immunoprecipitation) from other molecules in a cell lysate, in Western blot analyses to identify proteins separated by electrophoresis, and in immunohistochemistry or immunofluorescence to examine protein expression in tissue sections or to locate proteins within cells with the assistance of a microscope. Proteins can also be detected and quantified with antibodies, using ELISA and ELISPOT techniques.

Structure Prediction

The importance of antibodies in health care and the biotechnology industry demands knowledge of their structures at high resolution. This information is used for protein engineering, modifying the antigen binding affinity, and identifying an epitope, of a given antibody. X-ray crystallography is one commonly used method for determining antibody structures.

However, crystallizing an antibody is often laborious and time-consuming. Computational approaches provide a cheaper and faster alternative to crystallography, but their results are more equivocal, since they do not produce empirical structures. Online web servers such as Web Antibody Modelling (WAM) and Prediction of Immunoglobulin Structure (PIGS) enables computational modelling of antibody variable regions. Rosetta Antibody is a novel antibody F_V region structure prediction server, which incorporates sophisticated techniques to minimize CDR loops and optimize the relative orientation of the light and heavy chains, as well as homology models that predict successful docking of antibodies with their unique antigen.

History

The first use of the term "antibody" occurred in a text by Paul Ehrlich. The term Antikörper (the German word for antibody) appears in the conclusion of his article "Experimental Studies on Immunity", published in October 1891, which states that, "if two substances give rise to two different antikörper, then they themselves must be different". However, the term was not accepted immediately and several other terms for antibody were proposed; these included Immunkörper, Amboceptor, Zwischenkörper, substance sensibilisatrice, copula, Desmon, philocytase, fixateur, and Immunisin. The word antibody has formal analogy to the word antitoxin and a similar concept to Immunkörper.

The study of antibodies began in 1890 when Kitasato Shibasaburô described antibody activity against diphtheria and tetanus toxins. Kitasato put forward the theory of humoral immunity, proposing that a mediator in serum could react with a foreign antigen. His idea prompted Paul Ehrlich to propose the side-chain theory for antibody and antigen interaction in 1897, when he hypothesized that receptors (described as "side-chains") on the surface of cells could bind specifically to toxins – in a "lock-and-key" interaction – and that this binding reaction is the trigger for the production of antibodies. Other researchers believed that antibodies existed freely in the blood and, in 1904, Almroth Wright suggested that soluble antibodies coated bacteria to label them for phagocytosis and killing; a process that he named opsoninization.

In the 1920s, Michael Heidelberger and Oswald Avery observed that antigens could be precipitated by antibodies and went on to show that antibodies are made of protein. The biochemical properties of antigen-antibody-binding interactions were examined in more detail in the late 1930s by John Marrack. The next major advance was in the 1940s, when Linus Pauling confirmed the lock-and-key theory proposed by Ehrlich by showing that the interactions between antibodies and antigens depend more on their shape than their chemical composition. In 1948, Astrid Fagreaus discovered that B cells, in the form of plasma cells, were responsible for generating antibodies.

Further work concentrated on characterizing the structures of the antibody proteins. A major advance in these structural studies was the discovery in the early 1960s by Gerald Edelman and Joseph Gally of the antibody light chain, and their realisation that this protein is the same as the Bence-Jones protein described in 1845 by Henry Bence Jones.

Edelman went on to discover that antibodies are composed of disulfide bond-linked heavy and light chains. Around the same time, antibody-binding (Fab) and antibody tail (Fc) regions of IgG were characterized by Rodney Porter. Together, these scientists deduced the structure and complete amino acid sequence of IgG, a feat for which they were jointly awarded the 1972 Nobel Prize in Physiology or Medicine. The Fv fragment was prepared and characterized by David Givol. While most of these early studies focused on IgM and IgG, other immunoglobulin isotypes were identified in the 1960s: Thomas Tomasi discovered secretory antibody (IgA) and David S. Rowe and John L. Fahey identified IgD, and IgE was identified by Kimishige Ishizaka and Teruko Ishizaka as a class of antibodies involved in allergic reactions. In a landmark series of experiments beginning in 1976, Susumu Tonegawa showed that genetic material can rearrange itself to form the vast array of available antibodies.

Biomagnetism

Biomagnetism is the phenomenon of magnetic fields *produced* by living organisms; it is a subset of bioelectromagnetism. In contrast, organisms' use of magnetism in navigation is magnetoception and the study of the magnetic fields' *effects* on organisms is *magnetobiology*. (The word biomagnetism has also been used loosely to include magnetobiology, further encompassing almost any combination of the words magnetism, cosmology, and biology, such as "magnetoastrobiology".)

The origin of the word biomagnetism is unclear, but seems to have appeared several hundred years ago, linked to the expression "animal magnetism." The present scientific definition took form in the 1970s, when an increasing number of researchers began to measure the magnetic fields produced by the human body. The first valid measurement was actually made in 1963, but the field began to expand only after a low-noise technique was developed in 1970. Today the community of biomagnetic researchers does not have a formal organisation, but international conferences are held every two years, with about 600 attendees. Most conference activity centres on the MEG (magnetoencephalogram), the measurement of the magnetic field of the brain.

In 2013 a study published in the journal Frontiers in Zoology found that "dogs preferred to excrete with the body being aligned along the North–south axis under calm MF conditions", and that these findings "open new horizons for biomagnetic research."

7

Carbon Nanotube and Fullerene Sensors

A fullerene is any molecule composed entirely of carbon, in the form of a hollow sphere, ellipsoid, tube, and many other shapes. Spherical fullerenes are also called buckyballs, and they resemble the balls used in football (soccer). Cylindrical ones are called carbon nanotubes or buckytubes. Fullerenes are similar in structure to graphite, which is composed of stacked graphene sheets of linked hexagonal rings; but they may also contain pentagonal (or sometimes heptagonal) rings.

The first fullerene molecule to be discovered, and the family's namesake, buckminsterfullerene (C_{60}), was prepared in 1985 by Richard Smalley, Robert Curl, James Heath, Sean O'Brien, and Harold Kroto at Rice University. The name was a homage to Buckminster Fuller, whose geodesic domes it resembles. The structure was also identified some five years earlier by Sumio Iijima, from an electron microscope image, where it formed the core of a "bucky onion." Fullerenes have since been found to occur in nature. More recently, fullerenes have been detected in outer space. According to astronomer Letizia Stanghellini, "It's possible that buckyballs from outer space provided seeds for life on Earth."

The discovery of fullerenes greatly expanded the number of known carbon allotropes, which until recently were limited to graphite, diamond, and amorphous carbon such as soot and charcoal. Buckyballs and buckytubes have been the subject of intense research, both for their unique chemistry and for their technological applications, especially in materials science, electronics, and nanotechnology.

History

The icosahedral $C_{60}H_{60}$ cage was mentioned in 1965 as a possible topological structure. Eiji Osawa of Toyohashi University of Technology predicted the existence of C_{60} in 1970. He noticed that the structure of a corannulene molecule was a subset of football shape, and he hypothesised that a full ball shape could also exist. Japanese scientific journals reported his idea, but it did not reach Europe or the Americas.

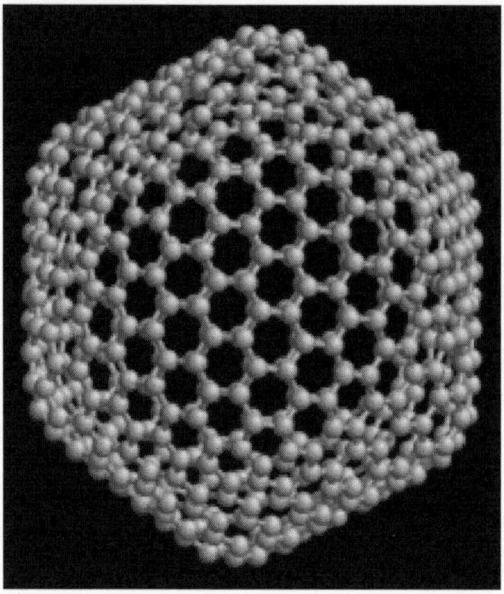

Figure: *The icosahedral fullerene C_{540}, another member of the family of fullerenes.*

Also in 1970, R. W. Henson (then of the Atomic Energy Research Establishment) proposed the structure and made a model of C_{60}. Unfortunately, the evidence for this new form of carbon was very weak and was not accepted, even by his colleagues. The results were never published but were acknowledged in *Carbon* in 1999.

Independently from Henson, in 1973 a group of scientists from the USSR, directed by Prof. Bochvar, made a quantum-chemical analysis of the stability of C_{60} and calculated its electronic structure. As in the previous cases, the scientific community did not accept the theoretical prediction. The paper was published in 1973 in *Proceedings of the USSR Academy of Sciences* (in Russian).

In mass spectrometry, discrete peaks appeared corresponding to molecules with the exact mass of sixty or seventy or more carbon atoms. In 1985, Harold Kroto (then of the University of Sussex), James R. Heath, Sean O'Brien, Robert Curl and Richard Smalley, from Rice University, discovered C_{60}, and shortly thereafter came to discover the fullerenes. Kroto, Curl, and Smalley were awarded the 1996 Nobel Prize in Chemistry for their roles in the discovery of this class of molecules. C_{60} and other fullerenes were later noticed occurring outside the laboratory (for example, in normal candle-soot).

By 1991, it was relatively easy to produce gram-sized samples of fullerene powder using the techniques of Donald Huffman, Wolfgang Krdtschmer and Konstantinos Fostiropoulos. Fullerene purification remains a challenge to chemists and to a large extent determines fullerene prices. So-called endohedral fullerenes have ions or small molecules incorporated inside the cage atoms. Fullerene is an unusual reactant in many organic reactions such as the Bingel reaction discovered in 1993. Carbon nanotubes were recognised in 1991.

Minute quantities of the fullerenes, in the form of C_{60}, C_{70}, C_{76}, C_{82} and C_{84} molecules, are produced in nature, hidden in soot and formed by lightning discharges in the atmosphere. In 1992, fullerenes were found in a family of minerals known as Shungites in Karelia, Russia.

In 2010, fullerenes (C_{60}) have been discovered in a cloud of cosmic dust surrounding a distant star 6500 light years away. Using NASA's Spitzer infrared telescope the scientists spotted the molecules' unmistakable infrared signature. Sir Harry Kroto, who shared the 1996 Nobel Prize in Chemistry for the discovery of buckyballs commented: "This most exciting breakthrough provides convincing evidence that the buckyball has, as I long suspected, existed since time immemorial in the dark recesses of our galaxy."

Naming

Buckminsterfullerene (C_{60}) was named after Richard Buckminster Fuller, a noted architectural modeler who popularized the geodesic dome. Since buckminsterfullerenes have a shape similar to that sort of dome, the name was thought appropriate. As the discovery of the fullerene family came *after* buckminsterfullerene, the shortened name 'fullerene' is used to refer to the family of fullerenes. The suffix "-ene" indicates that each C atom is covalently bonded to three others (instead of the maximum of four), a situation that classically would correspond to the existence of bonds involving two pairs of electrons ("double bonds").

Types of Fullerene

Since the discovery of fullerenes in 1985, structural variations on fullerenes have evolved well beyond the individual clusters themselves. Examples include:

- *Buckyball clusters:* smallest member is C_{20} (unsaturated version of dodecahedrane) and the most common is C_{60};
- *Nanotubes:* hollow tubes of very small dimensions, having single or multiple walls; potential applications in electronics industry;

- *Megatubes:* larger in diameter than nanotubes and prepared with walls of different thickness; potentially used for the transport of a variety of molecules of different sizes;
- *Polymers:* chain, two-dimensional and three-dimensional polymers are formed under high-pressure high-temperature conditions; single-strand polymers are formed using the Atom Transfer Radical Addition Polymerization (ATRAP) route;
- *Nano"onions":* spherical particles based on multiple carbon layers surrounding a buckyball core; proposed for lubricants;
- linked "ball-and-chain" dimers: two buckyballs linked by a carbon chain;
- fullerene rings.

Buckyballs

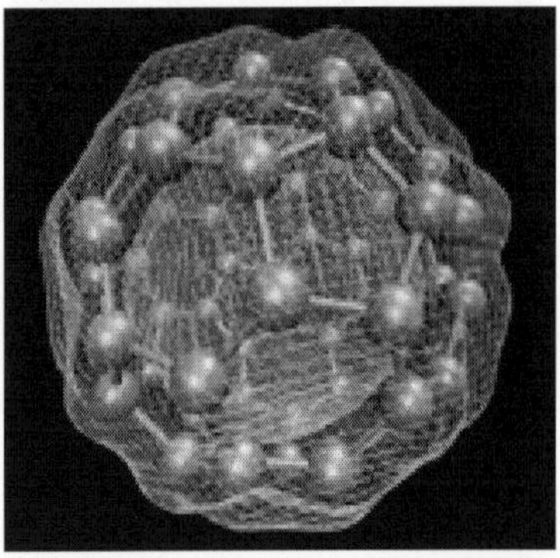

Figure: *C_{60} with isosurface of ground state electron density as calculated with DFT*

Figure: *Many association footballs have the same shape as buckminsterfullerene, C_{60}*

Buckminsterfullerene

Buckminsterfullerene is the smallest fullerene molecule containing pentagonal and hexagonal rings in which no two pentagons share an edge (which can be destabilizing, as in pentalene). It is also the most common in terms of natural occurrence, as it can often be found in soot.

The structure of C_{60} is a truncated icosahedron, which resembles an association football ball of the type made of twenty hexagons and twelve pentagons, with a carbon atom at the vertices of each polygon and a bond along each polygon edge.

The van der Waals diameter of a C_{60} molecule is about 1.1 nanometres (nm). The nucleus to nucleus diameter of a C_{60} molecule is about 0.71 nm.

The C_{60} molecule has two bond lengths. The 6:6 ring bonds (between two hexagons) can be considered "double bonds" and are shorter than the 6:5 bonds (between a hexagon and a pentagon). Its average bond length is 1.4 angstroms.

Silicon buckyballs have been created around metal ions.

Boron Buckyball

A type of buckyball which uses boron atoms, instead of the usual carbon, was predicted and described in 2007. The B_{80} structure, with each atom forming 5 or 6 bonds, is predicted to be more stable than the C_{60} buckyball. One reason for this given by the researchers is that the B-80 is actually more like the original geodesic dome structure popularized by Buckminster Fuller, which uses triangles rather than hexagons. However, this work has been subject to much criticism by quantum chemists as it was concluded that the predicted I_h symmetric structure was vibrationally unstable and the resulting cage undergoes a spontaneous symmetry break, yielding a puckered cage with rare T_h symmetry (symmetry of a volleyball). The number of six-member rings in this molecule is 20 and number of five-member rings is 12. There is an additional atom in the centre of each six-member ring, bonded to each atom surrounding it. By employing a systematic global search algorithm, later it was found that the previously proposed B80 fullerene is not global minimum for 80 atom boron clusters and hence can not be found in nature. In the same paper by Sandip De et al, it was concluded that born energy land scape is significantly different than other fullerenes already found in nature hence pure boron fullerenes are unlikely to exist in nature.

Other Buckyballs

Another fairly common fullerene is C_{70}, but fullerenes with 72, 76, 84 and even up to 100 carbon atoms are commonly obtained.

In mathematical terms, the structure of a fullerene is a trivalent convex polyhedron with pentagonal and hexagonal faces. In graph theory, the term fullerene refers to any 3-regular, planar graph with all faces of size 5 or 6 (including the external face). It follows from Euler's polyhedron formula, $V -$

$E + F = 2$ (where V, E, F are the numbers of vertices, edges, and faces), that there are exactly 12 pentagons in a fullerene and $V/2 - 10$ hexagons.

The smallest fullerene is the dodecahedral C_{20}. There are no fullerenes with 22 vertices. The number of fullerenes C_{2n} grows with increasing $n = 12, 13, 14, ...$, roughly in proportion to n^2 (sequence A007894 in OEIS). For instance, there are 1812 non-isomorphic fullerenes C_{60}. Note that only one form of C_{60}, the buckminsterfullerene alias truncated icosahedron, has no pair of adjacent pentagons (the smallest such fullerene). To further illustrate the growth, there are 214,127,713 non-isomorphic fullerenes C_{200}, 15,655,672 of which have no adjacent pentagons. Optimized structures of many fullerene isomers are published and listed on the web.

Trimetasphere carbon nanomaterials were discovered by researchers at Virginia Tech and licensed exclusively to Luna Innovations. This class of novel molecules comprises 80 carbon atoms (C_{80}) forming a sphere which encloses a complex of three metal atoms and one nitrogen atom. These fullerenes encapsulate metals which puts them in the subset referred to as metallofullerenes. Trimetaspheres have the potential for use in diagnostics (as safe imaging agents), therapeutics and in organic solar cells.

Carbon Nanotubes

Nanotubes are cylindrical fullerenes. These tubes of carbon are usually only a few nanometres wide, but they can range from less than a micrometre to several millimetres in length. They often have closed ends, but can be open-ended as well. There are also cases in which the tube reduces in diameter before closing off. Their unique molecular structure results in extraordinary macroscopic properties, including high tensile strength, high electrical conductivity, high ductility, high heat conductivity, and relative chemical inactivity (as it is cylindrical and "planar" — that is, it has no "exposed" atoms that can be easily displaced). One proposed use of carbon nanotubes is in paper batteries, developed in 2007 by researchers at Rensselaer Polytechnic Institute. Another highly speculative proposed use in the field of space technologies is to produce high-tensile carbon cables required by a space elevator.

Carbon Nanobuds

Nanobuds have been obtained by adding buckminsterfullerenes to carbon nanotubes.

Fullerite

Fullerites are the solid-state manifestation of fullerenes and related compounds and materials.

"Ultrahard fullerite" is a coined term frequently used to describe material produced by high-pressure high-temperature (HPHT) processing of fullerite. Such treatment converts fullerite into a nanocrystalline form of diamond which has been reported to exhibit remarkable mechanical properties.

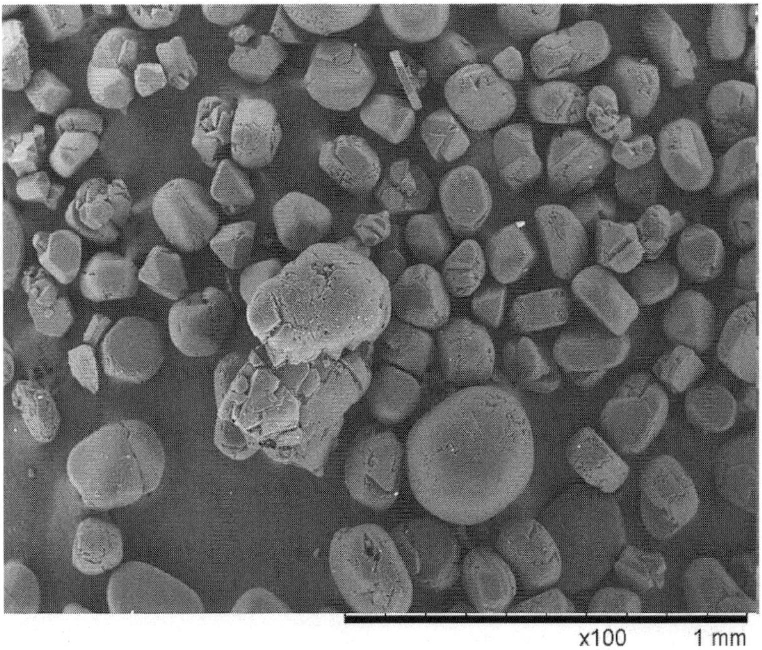

Figure: *Fullerite (scanning electron microscope image)*

Inorganic Fullerenes

Materials with fullerene-like molecular structures but lacking carbon include MoS_2, WS_2, TiS_2 and NbS_2. Prof. J. M. Martin from Ecole Central de Lyon in France tested the new material under isostatic pressure and found it to be stable up to at least 350 tons/cm^2.

Properties

For the past decade, the chemical and physical properties of fullerenes have been a hot topic in the field of research and development, and are likely to continue to be for a long time. Popular Science has discussed possible uses of fullerenes (graphene) in armor. In April 2003, fullerenes were under study for potential medicinal use: binding specific antibiotics to the structure to target resistant bacteria and even target certain cancer cells such as melanoma. The October 2005 issue of Chemistry & Biology contains an article describing the use of fullerenes as light-activated antimicrobial agents.

In the field of nanotechnology, heat resistance and superconductivity are some of the more heavily studied properties.

A common method used to produce fullerenes is to send a large current between two nearby graphite electrodes in an inert atmosphere. The resulting carbon plasma arc between the electrodes cools into sooty residue from which many fullerenes can be isolated.

There are many calculations that have been done using ab-initio quantum methods applied to fullerenes. By DFT and TD-DFT methods one can obtain IR, Raman and UV spectra. Results of such calculations can be compared with experimental results.

Aromaticity

Researchers have been able to increase the reactivity of fullerenes by attaching active groups to their surfaces. Buckminsterfullerene does not exhibit "superaromaticity": that is, the electrons in the hexagonal rings do not delocalize over the whole molecule.

A spherical fullerene of n carbon atoms has n pi-bonding electrons, free to delocalize. These should try to delocalize over the whole molecule. The quantum mechanics of such an arrangement should be like one shell only of the well-known quantum mechanical structure of a single atom, with a stable filled shell for $n = 2, 8, 18, 32, 50, 72, 98, 128$, etc.; i.e. twice a perfect square number; but this series does not include 60. This $2(N + 1)^2$ rule (with N integer) for spherical aromaticity is the three-dimensional analogue of Hückel's rule. The 10+ cation would satisfy this rule, and should be aromatic. This has been shown to be the case using quantum chemical modelling, which showed the existence of strong diamagnetic sphere currents in the cation.

As a result, C_{60} in water tends to pick up two more electrons and become an anion. The nC_{60} described below may be the result of C_{60} trying to form a loose metallic bond.

Chemistry

Fullerenes are stable, but not totally unreactive. The sp^2-hybridized carbon atoms, which are at their energy minimum in planar graphite, must be bent to form the closed sphere or tube, which produces angle strain. The characteristic reaction of fullerenes is electrophilic addition at 6,6-double bonds, which reduces angle strain by changing sp^2-hybridized carbons into sp^3-hybridized ones. The change in hybridized orbitals causes the bond angles to decrease from about 120° in the sp^2 orbitals to about 109.5° in the sp^3 orbitals. This decrease in bond angles allows for the bonds to bend less when closing the sphere or tube, and thus, the molecule becomes more stable.

Other atoms can be trapped inside fullerenes to form inclusion compounds known as endohedral fullerenes. An unusual example is the egg shaped fullerene $Tb_3N@C_{84}$, which violates the isolated pentagon rule. Recent evidence for a meteor impact at the end of the Permian period was found by analyzing noble gases so preserved. Metallofullerene-based inoculates using the rhonditic steel process are beginning production as one of the first commercially-viable uses of buckyballs.

Solubility

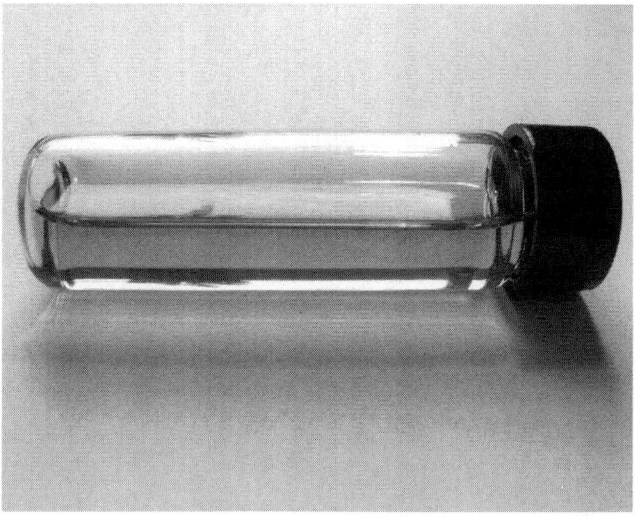

Figure: *C_{60} in solution*

Fullerenes are sparingly soluble in many solvents. Common solvents for the fullerenes include aromatics, such as toluene, and others like carbon disulfide. Solutions of pure buckminsterfullerene have a deep purple colour. Solutions of C_{70} are a reddish brown.

C_{60} in extra virgin olive oil showing the characteristic purple colour of pristine C_{60} solutions.

The higher fullerenes C_{76} to C_{84} have a variety of colours. C_{76} has two optical forms, while other higher fullerenes have several structural isomers. Fullerenes are the only known allotrope of carbon that can be dissolved in common solvents at room temperature.

Solvent	C_{60}	C_{70}
1-chloronaphthalene	51 mg/mL	*
1-methylnaphthalene	33 mg/mL	*
1,2-dichlorobenzene	24 mg/mL	36.2 mg/mL
1,2,4-trimethylbenzene	18 mg/mL	*
tetrahydronaphthalene	16 mg/mL	*
carbon disulfide	8 mg/mL	9.875 mg/mL
1,2,3-tribromopropane	8 mg/mL	*
chlorobenzene	7 mg/mL	*
xylene	5 mg/mL	3.985 mg/mL(p-xylene)
bromoform	5 mg/mL	*
cumene	4 mg/mL	*
toluene	3 mg/mL	1.406 mg/mL
benzene	1.5 mg/mL	1.3 mg/mL
carbon tetrachloride	0.447 mg/mL	0.121 mg/mL
chloroform	0.25 mg/mL	*
n-hexane	0.046 mg/mL	0.013 mg/mL
cyclohexane	0.035 mg/mL	0.08 mg/mL
tetrahydrofuran	0.006 mg/mL	*
acetonitrile	0.004 mg/mL	*
methanol	0.000 04 mg/mL	*
water	1.3×10^{-11} mg/mL	*
pentane	0.004 mg/mL	0.002 mg/mL
heptane	*	0.047 mg/mL
octane	0.025 mg/mL	0.042 mg/mL
isooctane	0.026 mg/mL	*
decane	0.070 mg/mL	0.053 mg/mL
dodecane	0.091 mg/mL	0.098 mg/mL
tetradecane	0.126 mg/mL	*
acetone	*	0.0019 mg/mL
isopropanol	*	0.0021 mg/mL
dioxane	0.0041 mg/mL	*
mesitylene	0.997 mg/mL	1.472 mg/mL
dichloromethane	0.254 mg/mL	0.080 mg/mL

* *: Solubility not measured*

Some fullerene structures are not soluble because they have a small band gap between the ground and excited states. These include the small fullerenes C_{28}, C_{36} and C_{50}. The C_{72} structure is also in this class, but the endohedral version with a trapped lanthanide-group atom is soluble due to the interaction of the metal atom and the electronic states of the fullerene. Researchers had originally been puzzled by C_{72} being absent in fullerene plasma-generated soot extract, but found in endohedral samples. Small band gap fullerenes are highly reactive and bind to other fullerenes or to soot particles.

Solvents that are able to dissolve buckminsterfullerene (C_{60} and C_{70}) are listed at left in order from highest solubility. The solubility value given is the approximate saturated concentration.

Solubility of C_{60} in some solvents shows unusual behaviour due to existence of solvate phases (analogues of crystallohydrates). For example, solubility of C_{60} in benzene solution shows maximum at about 313 K. Crystallization from benzene solution at temperatures below maximum results in formation of triclinic solid solvate with four benzene molecules $C_{60} \cdot 4C_6H_6$ which is rather unstable in air. Out of solution, this structure decomposes into usual face-centred cubic (fcc) C_{60} in few minutes' time. At temperatures above solubility maximum the solvate is not stable even when immersed in saturated solution and melts with formation of fcc C_{60}. Crystallization at temperatures above the solubility maximum results in formation of pure fcc C_{60}. Millimetre-sized crystals of C_{60} and C_{70} can be grown from solution both for solvates and for pure fullerenes.

Quantum Mechanics

In 1999, researchers from the University of Vienna demonstrated that wave-particle duality applied to molecules such as fullerene. One of the co-authors of this research, Julian Voss-Andreae, has since created several sculptures symbolizing wave-particle duality in fullerenes.

Superconductivity

Chirality: Some fullerenes (e.g. C_{76}, C_{78}, C_{80}, and C_{84}) are inherently chiral because they are D_2-symmetric, and have been successfully resolved. Research efforts are ongoing to develop specific sensors for their enantiomers.

Construction

Two theories have been proposed to describe the molecular mechanisms that make fullerenes. The older, "bottom-up" theory proposes that they are built atom-by-atom. The alternative "top-down" approach claims that fullerenes form when much larger structures break into constituent parts.

In 2013 researchers discovered that asymmetrical fullerenes formed from larger structures settle into stable fullerenes. The synthesized substance was a particular metallofullerene consisting of 84 carbon atoms with two additional carbon atoms and two yttrium atoms inside the cage. The process produced approximately 100 micrograms.

However, they found that the asymmetrical molecule could theoretically collapse to form nearly every known fullerene and metallofullerene. Minor perturbations involving the breaking of a few molecular bonds cause the cage to become highly symmetrical and stable. This insight supports the theory that fullerenes can be formed from graphene when the appropriate molecular bonds are severed.

Applications

Tumor Research: While past cancer research has involved radiation therapy, photodynamic therapy is important to study because breakthroughs in treatments for tumor cells will give more options to patients with different conditions. More recent experiments using HeLa cells in cancer research involves the development of new photosensitizers with increased ability to be absorbed by cancer cells and still trigger cell death. It is also important that a new photosensitizer does not stay in the body for a long time to prevent unwanted cell damage.

Fullerenes can be made to be absorbed by HeLa cells. The C_{60} derivatives can be delivered to the cells by using the functional groups L-phenylalanine, folic acid, and L-arginine among others. The purpose for functionalizing the fullerenes is to increase the solubility of the molecule by the cancer cells. Cancer cells take up these molecules at an increased rate because of an upregulation of transporters in the cancer cell, in this case amino acid transporters will bring in the L-arginine and L-phenylalanine functional groups of the fullerenes.

Once absorbed by the cells, the C_{60} derivatives would react to light radiation by turning molecular oxygen into reactive oxygen which triggers apoptosis in the HeLa cells and other cancer cells that can absorb the fullerene molecule. This research shows that a reactive substance can target cancer cells and then be triggered by light radiation, minimizing damage to surrounding tissues while undergoing treatment.

When absorbed by cancer cells and exposed to light radiation, the reaction that creates reactive oxygen damages the DNA, proteins, and lipids that make up the cancer cell. This cellular damage forces the cancerous cell to go through apoptosis, which can lead to the reduction in size of a tumor. Once the light radiation treatment is finished the fullerene will reabsorb the free radicals to prevent damage of other tissues. Since this treatment focuses on cancer cells it is a good option for patients whose cancer cells are within reach of light radiation. As this research continues into the future it will be able to penetrate deeper into the body and more effectively absorbed by cancer cells.

Safety and Toxicity

Moussa et al. (1996–7) studied the in vivo toxicity of C_{60} after intra-peritoneal administration of large doses. No evidence of toxicity was found and the mice tolerated a dose of 5 g/kg of body weight. Mori et al. (2006) could not find toxicity in rodents for C_{60} and C_{70} mixtures after oral administration of a dose of 2 g/kg body weight and did not observe evidence of genotoxic or mutagenic potential in vitro. Other studies could not establish the toxicity of fullerenes: on the contrary, the work of Gharbi et al. (2005) suggested that aqueous C_{60} suspensions failing to produce acute or subacute toxicity in rodents could also protect their livers in a dose-dependent manner against free-radical damage. In 2012, it was revealed that oral administration prolonged the lifespan of rats by nearly 100% without showing toxic effects. In a more recent video interview filmed in October 2012 and placed online, Professor Moussa states that pure C_{60} is not toxic.

A comprehensive and recent review on fullerene toxicity is given by Kolosnjaj et al. (2007). These authors review the works on fullerene toxicity beginning in the early 1990s to present, and conclude that very little evidence gathered since the discovery of fullerenes indicate that C_{60} is toxic.

With reference to nanotubes, a 2008 study on carbon nanotubes introduced into the abdominal cavity of mice led the authors to suggest comparisons to "asbestos-like pathogenicity". It should be noted that this was not an inhalation study, though there have been several performed in the past, therefore it is premature to conclude that nanotubes should be considered to have a toxicological profile similar to asbestos. Conversely, and perhaps illustrative of how the various classes of molecules which fall under the general term fullerene cover a wide range of properties, Sayes et al. found that in vivo inhalation of $C_{60}(OH)_{24}$ and nano-C_{60} in rats gave no effect, whereas in comparison quartz particles produced an inflammatory response under the same conditions. As stated above, nanotubes are quite different in chemical and physical properties to C_{60}, i.e., molecular weight, shape, size, physical properties (such as solubility) all are very different, so from a toxicological standpoint, different results for C_{60} and nanotubes are not suggestive of any discrepancy in the findings.

When considering toxicological data, care must be taken to distinguish as necessary between what are normally referred to as fullerenes: (C_{60}, C_{70}, ...); fullerene derivatives: C_{60} or other fullerenes with covalently bonded chemical groups; fullerene complexes (e.g., water-solubilized with surfactants, such as

C_{60}-PVP; host-guest complexes, such as with cyclodextrin), where the fullerene is supermolecular bound to another molecule; C_{60} nanoparticles, which are extended solid-phase aggregates of C_{60} crystallites; and nanotubes, which are generally much larger (in terms of molecular weight and size) molecules, and are different in shape to the spheroidal fullerenes C_{60} and C_{70}, as well as having different chemical and physical properties.

The above different molecules span the range from insoluble materials in either hydrophilic or lipophilic media, to hydrophilic, lipophilic, or even amphiphilic molecules, and with other varying physical and chemical properties. Therefore any broad generalization extrapolating for example results from C_{60} to nanotubes or vice versa is not possible, though technically all are fullerenes, as the term is defined as a close-caged all-carbon molecule.

Any extrapolation of results from one molecule to other molecules must take into account considerations based on a quantitative structural analysis relationship study (QSARS), which mostly depends on how close the molecules under consideration are in physical and chemical properties.

Hydrogen Sensors Based on TiO_2

A hydrogen sensor is a gas detector that detects the presence of hydrogen. They contain micro-fabricated point-contact hydrogen sensors and are used to locate leaks. They are considered low-cost, compact, durable, and easy to maintain as compared to conventional gas detecting instruments.

Key Issues

There are five key issues with hydrogen detectors:

- Reliability: Functionality should be easily verifiable.
- Performance: Detection 0.5% hydrogen in air or better
- Response time < 1 second.
- Lifetime: At least the time between scheduled maintenance.
- Cost: Goal is $5 per sensor and $30 per controller.

Additional Requirements

- Measurement range coverage of 0.1%–10.0% concentration
- Operation in temperatures of -30°C to 80°C
- Accuracy within 5% of full scale
- Function in an ambient air gas environment within a 10%–98% relative humidity range
- Resistance to hydrocarbon and other interference.
- Lifetime greater than 10 years

Types of Microsensors

There are various types of hydrogen microsensors, which use different mechanisms to detect the gas. Palladium is used in many of these, because it selectively absorbs hydrogen gas and forms the compound palladium hydride. Palladium-based sensors have a strong temperature dependence which makes their response time too large at very low temperatures. Palladium sensors have to be protected against carbon monoxide, sulfur dioxide and hydrogen sulfide.

Optical Fibre Hydrogen Sensors

Several types of optical fibre surface plasmon resonance (SPR) sensor are used for the point-contact detection of hydrogen:

- Fiber Bragg grating coated with a palladium layer - Detects the hydrogen by metal hindrance.
- Micromirror - With a palladium thin layer at the cleaved end, detecting changes in the backreflected light.
- Tapered fibre coated with palladium - Hydrogen changes the refractive index of the palladium, and consequently the amount of losses in the evanescent wave.

Other Types of Hydrogen Sensors

- Electrochemical hydrogen sensor - low (ppm) levels of hydrogen gas can be sensed using electrochemical sensors which comprise an array of electrodes packaged so as to be surrounded by a conductive electrolyte and gas ingress controlled with a diffusion limited capillary.
- MEMS hydrogen sensor - The combination of nanotechnology and microelectromechanical systems (MEMS) technology allows the production of a hydrogen microsensor that functions properly at room temperature. One type of MEMS-based hydrogen sensor is coated with a film consisting of nanostructured indium oxide (In_2O_3) and tin oxide (SnO_2). A typical configuration for mechanical Pd-based hydrogen sensors is the usage of a free-standing cantilever that is coated with Pd. In the presence of H_2, the Pd layer expands and thereby induces a stress that causes the cantilever to bend. Pd-coated nano-mechanical resonators have also been reported in literature, relying on the stress-induced mechanical resonance frequency shift caused by the presence of H_2 gas. In this case, the response speed was enhanced through the use of a very thin layer of Pd (20 nm). Moderate heating was presented as a solution to the response impairment observed in humid conditions.

- Thin film sensor - A palladium thin film sensor is based on an opposing property that depends on the nanoscale structures within the thin film. In the thin film, nanosized palladium particles swell when the hydride is formed, and in the process of expanding, some of them form new electrical connections with their neighbours. The resistance decreases because of the increased number of conducting pathways.
- Thick film sensor - Thick film hydrogen sensors rely on the fact that palladium hydride's electrical resistance is greater than the metal's resistance. The absorption of hydrogen causes a measurable increase in electrical resistance.
- Chemochromic hydrogen sensor - Reversible and irreversible chemochromic hydrogen sensors, a smart pigment paint that visually identifies hydrogen leaks by a change in colour. The sensor is also available as tape.
- Diode based Schottky sensor - A Schottky diode-based hydrogen gas sensor employs a palladium-alloy gate. Hydrogen can be selectively absorbed in the gate, lowering the Schottky energy barrier. A Pd/InGaP metal-semiconductor (MS) Schottky diode can detect a concentration of 15 parts per million (ppm) H_2 in air. Silicon carbide semiconductor or silicon substrates are used.
- Metallic La-Mg2-Ni which is electrical conductive, absorbs hydrogen near ambient conditions, forming the nonmetallic hydride LaMg2NiH7 an insulator.

Sensors are typically calibrated at the manufacturing factory and are valid for the service life of the unit.

Enhancement

Siloxane enhances the sensitivity and reaction time of hydrogen sensors. Detection of hydrogen levels as low as 25 ppm can be achieved; far below hydrogen's lower explosive limit of around 40,000 ppm.

Titanium Dioxide (TiO_2)

Titanium dioxide, also known as titanium(IV) oxide or titania, is the naturally occurring oxide of titanium, chemical formula TiO_2. When used as a pigment, it is called titanium white, Pigment White 6 (PW6), or CI 77891. Generally it is sourced from ilmenite, rutile and anatase. It has a wide range of applications, from paint to sunscreen to food colouring. When used as a food colouring, it has E number E171.

Nanotubes

Anatase can be converted by hydrothermal synthesis to delaminated anatase inorganic nanotubes and titanate nanoribbons which are of potential interest as catalytic supports and photocatalysts. In the synthesis, anatase is mixed with 10 M sodium hydroxide and heated at 130 °C for 72 hours. The reaction product is washed with dilute hydrochloric acid and heated at 400 °C for another 15 hours. The yield of nanotubes is quantitative and the tubes have an outer diameter of 10 to 20 nm and an inner diameter of 5 to 8 nm and have a length of 1 μm. A higher reaction temperature (170 °C) and less reaction volume gives the corresponding nanowires.

Another process for synthesizing TiO_2 is through anodization in an electrolytic solution. When anodized in a 0.5 weight percent HF solution for 20 minutes, well-aligned titanium oxide nanotube arrays can be fabricated an average tube diameter of 60 nm and length of 250 nm. Based on X-ray Diffraction, nanotubes grown through anodization are amorphous.

Applications

The most important application areas are paints and varnishes as well as paper and plastics, which account for about 80% of the world's titanium dioxide consumption. Other pigment applications such as printing inks, fibers, rubber, cosmetic products and foodstuffs account for another 8%. The rest is used in other applications, for instance the production of technical pure titanium, glass and glass ceramics, electrical ceramics, catalysts, electric conductors and chemical intermediates.

Photocatalyst

Titanium dioxide, particularly in the anatase form, is a photocatalyst under ultraviolet (UV) light. Recently it has been found that titanium dioxide, when spiked with nitrogen ions or doped with metal oxide like tungsten trioxide, is also a photocatalyst under either visible or UV light. The strong oxidative potential of the positive holes oxidizes water to create hydroxyl radicals. It can also oxidize oxygen or organic materials directly. Hence, in addition to its use as a pigment, titanium dioxide can be added to paints, cements, windows, tiles, or other products for its sterilizing, deodorizing and anti-fouling properties and is used as a hydrolysis catalyst. It is also used in dye-sensitized solar cells, which are a type of chemical solar cell (also known as a Graetzel cell).

The photocatalytic properties of titanium dioxide were discovered by Akira Fujishima in 1967 and published in 1972. The process on the surface of the titanium dioxide was called the Honda-Fujishima effect. Titanium dioxide has potential for use in energy production: as a photocatalyst, it can carry out hydrolysis; i.e., break water into hydrogen and oxygen. With the hydrogen collected, it could be used as a fuel. The efficiency of this process can be greatly improved by doping the oxide with carbon. Further efficiency and durability has been obtained by introducing disorder to the lattice structure of the surface layer of titanium dioxide nanocrystals, permitting infrared absorption.

In 1995 Fujishima and his group discovered the superhydrophilicity phenomenon for titanium dioxide coated glass exposed to sun light. This resulted in the development of self-cleaning glass and anti-fogging coatings.

TiO_2 incorporated into outdoor building materials, such as paving stones in noxer blocks or paints, can substantially reduce concentrations of airborne pollutants such as volatile organic compounds and nitrogen oxides.

A photocatalytic cement that uses titanium dioxide as a primary component, produced by Italcementi Group, was included in Time's Top 50 Inventions of 2008.

Attempts have been made to photocatalytically mineralize pollutants (to convert into CO_2 and H_2O) in waste water. TiO_2 offers great potential as an industrial technology for detoxification or remediation of wastewater due to several factors:

1. The process uses natural oxygen and sunlight and thus occurs under ambient conditions; it is wavelength selective and is accelerated by UV light.
2. The photocatalyst is inexpensive, readily available, non-toxic, chemically and mechanically stable, and has a high turnover.
3. The formation of photocyclized intermediate products, unlike direct photolysis techniques, is avoided.
4. Oxidation of the substrates to CO_2 is complete.
5. TiO_2 can be supported on suitable reactor substrates.

Electronic Data Storage Medium

In 2010, researchers at the University of Tokyo, Japan have created a crystal form of titanium oxide with particles 5 to 20 nanometres that can be switched between two states with light. Use of the 5 nm particles could theoretically lead to a 25 TB storage disc.

Other Applications

- Titanium dioxide in solution or suspension can be used to cleave protein that contains the amino acid proline at the site where proline is present.
- Titanium dioxide is also used as a material in the memristor, a new electronic circuit element. It can be employed for solar energy conversion based on dye, polymer, or quantum dot sensitized nanocrystalline TiO_2 solar cells using conjugated polymers as solid electrolytes.

- Synthetic single crystals and films of TiO_2 are used as a semiconductor, and also in Bragg-stack style dielectric mirrors due to the high refractive index of TiO_2 (2.5–2.9).

Quantum Dots for Sensing

A quantum dot is a nanocrystal made of semiconductor materials that are small enough to exhibit quantum mechanical properties. Specifically, its excitons are confined in all three spatial dimensions.

The electronic properties of these materials are intermediate between those of bulk semiconductors and of discrete molecules. Quantum dots were discovered in a glass matrix by Alexei Ekimov and in colloidal solutions by Louis E. Brus. The term "quantum dot" was coined by Mark Reed.

Researchers have studied applications for quantum dots in transistors, solar cells, LEDs, and diode lasers. They have also investigated quantum dots as agents for medical imaging and as possible qubits in quantum computing.

Electronic characteristics of a quantum dot are closely related to its size and shape. For example, the band gap in a quantum dot which determines the frequency range of emitted light is inversely related to its size.

In fluorescent dye applications the frequency of emitted light increases as the size of the quantum dot decreases. Consequently, the colour of emitted light shifts from red to blue when the size of the quantum dot is made smaller.

This allows the excitation and emission of quantum dots to be highly tunable. Since the size of a quantum dot may be set when it is made, its conductive properties may be carefully controlled. Quantum dot assemblies consisting of many different sizes, such as gradient multi-layer nanofilms, can be made to exhibit a range of desirable emission properties.

Quantum Confinement in Semiconductors

In a semiconductor crystallite whose diameter is smaller than the size of its exciton Bohr radius, the excitons are squeezed, leading to quantum confinement.

The energy levels can then be modelled using the particle in a box model in which the energy of different states is dependent on the length of the box. Quantum dots are said to be in the 'weak confinement regime' if their radii are on the order of the exciton Bohr radius; quantum dots are said to be in the 'strong confinement regime' if their radii are smaller than the exciton Bohr radius. If the size of the quantum dot is small enough that the quantum confinement effects dominate (typically less than 10 nm), the electronic and optical properties are highly tunable.

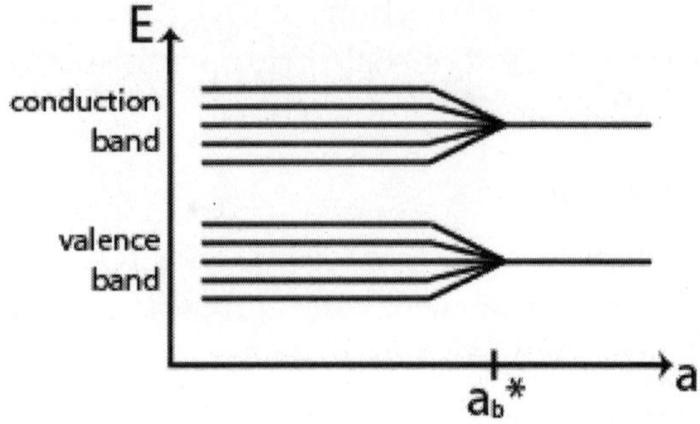

Figure: *Splitting of energy levels for small quantum dots due to the quantum confinement effect. The horizontal axis is the radius, or the size, of the quantum dots and* a_b** *is the Exciton Bohr radius.*

Fluorescence occurs when an excited electron relaxes to the ground state and combines with the hole. In a simplified model, the energy of the emitted photon can be understood as the sum of the band gap energy between the occupied level and the unoccupied energy level, the confinement energies of the hole and the excited electron, and the bound energy of the exciton (the electron-hole pair):

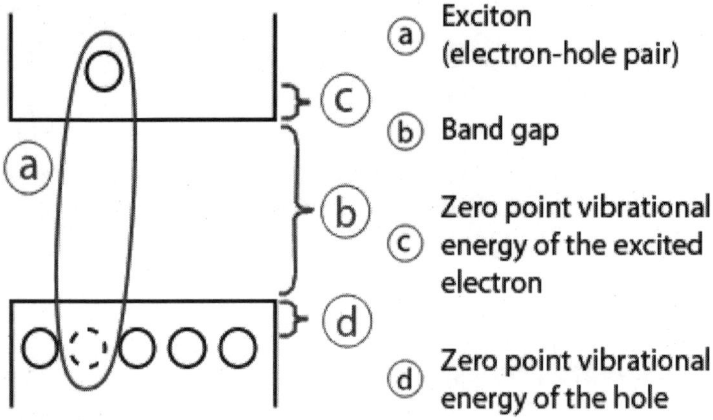

Band Gap Energy: The band gap can become larger in the strong confinement regime where the size of the quantum dot is smaller than the Exciton Bohr radius a_b^* as the energy levels split up.

$$a_b^* = \varepsilon_r \left(\frac{m}{\mu} \right) a_b$$

where a_b is the Bohr radius=0.053 nm, m is the mass, μ is the reduced mass, and ε_r is the size-dependent dielectric constant

This results in the increase in the total emission energy (the sum of the energy levels in the smaller band gaps in the strong confinement regime is larger than the energy levels in the band gaps of the original levels in the weak confinement regime) and the emission at various wavelengths; which is precisely what happens in the sun, where the quantum confinement effects are completely dominant and the energy levels split up to the degree that the energy spectrum is almost continuous, thus emitting white light.

Confinement Energy: The exciton entity can be modelled using the particle in the box. The electron and the hole can be seen as hydrogen in the Bohr model with the hydrogen nucleus replaced by the hole of positive charge and negative electron mass. Then the energy levels of the exciton can be represented as the solution to the particle in a box at the ground level (n = 1) with the mass replaced by the reduced mass. Thus by varying the size of the quantum dot, the confinement energy of the exciton can be controlled.

Bound Exciton Energy: There is Coulomb attraction between the negatively charged electron and the positively charged hole. The negative energy involved in the attraction is proportional to Rydberg's energy and inversely proportional to square of the size-dependent dielectric constant of the semiconductor. When the size of the semiconductor crystal is smaller than the Exciton Bohr radius, the Coulomb interaction must be modified to fit the situation.

Therefore, the sum of these energies can be represented as:

$$
\begin{aligned}
E_{\text{confinement}} &= \frac{\hbar^2\pi^2}{2a^2}\left(\frac{1}{m_e}+\frac{1}{m_h}\right) = \frac{\hbar^2\pi^2}{2\mu a^2} \\
E_{\text{exciton}} &= -\frac{1}{\epsilon_r^2}\frac{\mu}{m_e}R_y = -R_y^* \\
E &= E_{\text{bandgap}} + E_{\text{confinement}} + E_{\text{exciton}} \\
&= E_{\text{bandgap}} + \frac{\hbar^2\pi^2}{2\mu a^2} - R_y^*
\end{aligned}
$$

where μ is the reduced mass, a is the radius, m_e is the free electron mass, m_h is the hole mass, and ε_r is the size-dependent dielectric constant.

Although the above equations were derived using simplifying assumptions, the implications are clear; the energy of the quantum dots are dependent on their size due to the quantum confinement effects, which dominate below the critical size leading to changes in the optical properties. This effect of quantum confinement on the quantum dots have been experimentally verified and is a key feature of many emerging electronic structures.

Besides confinement in all three dimensions (i.e., a quantum dot), other quantum confined semiconductors include:

- Quantum wires, which confine electrons or holes in two spatial dimensions and allow free propagation in the third.

- Quantum wells, which confine electrons or holes in one dimension and allow free propagation in two dimensions.

Production

There are several ways to confine excitons in semiconductors, resulting in different methods to produce quantum dots. In general, quantum wires, wells and dots are grown by advanced epitaxial techniques in nanocrystals produced by chemical methods or by ion implantation, or in nanodevices made by state-of-the-art lithographic techniques.

Colloidal Synthesis

Colloidal semiconductor nanocrystals are synthesized from precursor compounds dissolved in solutions, much like traditional chemical processes. The synthesis of colloidal quantum dots is based on a three-component system composed of precursors, organic surfactants, and solvents. When heating a reaction medium to a sufficiently high temperature, the precursors chemically transform into monomers.

Once the monomers reach a high enough supersaturation level, the nanocrystal growth starts with a nucleation process. The temperature during the growth process is one of the critical factors in determining optimal conditions for the nanocrystal growth. It must be high enough to allow for rearrangement and annealing of atoms during the synthesis process while being low enough to promote crystal growth.

Another critical factor that has to be stringently controlled during nanocrystal growth is the monomer concentration. The growth process of nanocrystals can occur in two different regimes, "focusing" and "defocusing". At high monomer concentrations, the critical size (the size where nanocrystals neither grow nor shrink) is relatively small, resulting in growth of nearly all particles. In this regime, smaller particles grow faster than large ones (since larger crystals need more atoms to grow than small crystals) resulting in "focusing" of the size distribution to yield nearly monodisperse particles. The size focusing is optimal when the monomer concentration is kept such that the average nanocrystal size present is always slightly larger than the critical size. When the monomer concentration is depleted during growth, the critical size becomes larger than the average size present, and the distribution "defocuses" as a result of Ostwald ripening.

There are colloidal methods to produce many different semiconductors. Typical dots are made of binary alloys such as cadmium selenide, cadmium sulfide, indium arsenide, and indium phosphide. Dots may also be made from ternary alloys such as cadmium selenide sulfide. These quantum dots can contain as few as 100 to 100,000 atoms within the quantum dot volume, with a diameter of 10 to 50 atoms. This corresponds to about 2 to 10 nanometres, and at 10 nm in diameter, nearly 3 million quantum dots could be lined up end to end and fit within the width of a human thumb.

Large batches of quantum dots may be synthesized via colloidal synthesis. Due to this scalability and the convenience of benchtop conditions, colloidal synthetic methods are promising for commercial applications. It is acknowledged to be the least toxic of all the different forms of synthesis.

Fabrication

- Self-assembled quantum dots are typically between 5 and 50 nm in size. Quantum dots defined by lithographically patterned gate electrodes, or by etching on two-dimensional electron gases in semiconductor heterostructures can have lateral dimensions exceeding 100 nm.
- Some quantum dots are small regions of one material buried in another with a larger band gap. These can be so-called core–shell structures, e.g., with CdSe in the core and ZnS in the shell or from special forms of silica called ormosil.
- Quantum dots sometimes occur spontaneously in quantum well structures due to monolayer fluctuations in the well's thickness.
- Self-assembled quantum dots nucleate spontaneously under certain conditions during molecular beam epitaxy (MBE) and metallorganic vapor phase epitaxy (MOVPE), when a material is grown on a substrate to which it is not lattice matched. The resulting strain produces coherently strained islands on top of a two-dimensional wetting layer. This growth mode is known as Stranski–Krastanov growth. The islands can be subsequently buried to form the quantum dot. This fabrication method has potential for applications in quantum cryptography (i.e. single photon sources) and quantum computation. The main limitations of this method are the cost of fabrication and the lack of control over positioning of individual dots.
- Individual quantum dots can be created from two-dimensional electron or hole gases present in remotely doped quantum wells or semiconductor heterostructures called lateral quantum dots. The sample surface is coated with a thin layer of resist. A lateral pattern is then defined in the resist by electron beam lithography. This pattern can then be transferred to the electron or hole gas by etching, or by depositing metal electrodes (lift-off process) that allow the application of external voltages between the electron gas and the electrodes. Such quantum dots are

mainly of interest for experiments and applications involving electron or hole transport, i.e., an electrical current.

- The energy spectrum of a quantum dot can be engineered by controlling the geometrical size, shape, and the strength of the confinement potential. Also, in contrast to atoms, it is relatively easy to connect quantum dots by tunnel barriers to conducting leads, which allows the application of the techniques of tunneling spectroscopy for their investigation.

The quantum dot absorption features correspond to transitions between discrete, three-dimensional particle in a box states of the electron and the hole, both confined to the same nanometre-size box. These discrete transitions are reminiscent of atomic spectra and have resulted in quantum dots also being called *artificial atoms*.

- Confinement in quantum dots can also arise from electrostatic potentials (generated by external electrodes, doping, strain, or impurities).
- CMOS technology can be employed to fabricate silicon quantum dots. Ultra small (L=20 nm, W=20 nm) CMOS transistors behave as single electron quantum dots when operated at cryogenic temperature over a range of –269 °C (4 K) to about "258 °C (15 K). The transistor displays Coulomb blockade due to progressive charging of electrons one by one. The number of electrons confined in the channel is driven by the gate voltage, starting from an occupation of zero electrons, and it can be set to 1 or many.

Viral Assembly

Lee et al. (2002) reported using genetically engineered M13 bacteriophage viruses to create quantum dot biocomposite structures. As a background to this work, it has previously been shown that genetically engineered viruses can recognise specific semiconductor surfaces through the method of selection by combinatorial phage display.

Additionally, it is known that liquid crystalline structures of wild-type viruses (Fd, M13, and TMV) are adjustable by controlling the solution concentrations, solution ionic strength, and the external magnetic field applied to the solutions. Consequently, the specific recognition properties of the virus can be used to organise inorganic nanocrystals, forming ordered arrays over the length scale defined by liquid crystal formation. Using this information, Lee et al. (2000) were able to create self-assembled, highly oriented, self-supporting films from a phage and ZnS precursor solution. This system allowed them to vary both the length of bacteriophage and the type of inorganic material through genetic modification and selection.

Electrochemical Assembly

Highly ordered arrays of quantum dots may also be self-assembled by electrochemical techniques. A template is created by causing an ionic reaction at an electrolyte-metal interface which results in the spontaneous assembly of nanostructures, including quantum dots, onto the metal which is then used as a mask for mesa-etching these nanostructures on a chosen substrate.

Bulk-Manufacture

Quantum dot manufacturing relies on a process called "high temperature dual injection" which has been scaled by multiple companies for commercial applications that require large quantities (100's of kgs to tonnes) of quantum dots. This is a reproducible production method that can be applied to a wide range of quantum dot sizes and compositions.

The bonding in certain cadmium-free quantum dots, such as III-V-based quantum dots, is more covalent than that in II-VI materials, therefore it is more difficult to separate nanoparticle nucleation and growth via a high temperature dual injection synthesis.

An alternative method of quantum dot synthesis, the "molecular seeding" process, provides a reproducible route to the production of high quality quantum dots in large volumes. The process utilises identical molecules of a molecular cluster compound as the nucleation sites for nanoparticle growth, thus avoiding the need for a high temperature injection step. Particle growth is maintained by the periodic addition of precursors at moderate temperatures until the desired particle size is reached. The molecular seeding process is not limited to the production of cadmium-free quantum dots; for example, the process can be used to synthesise kilogram batches of high quality II-VI quantum dots in just a few hours.

Another approach for the mass production of colloidal quantum dots can be seen in the transfer of the well-known hot-injection methodology for the synthesis to a technical continuous flow system. The batch-to-batch variations arising from the needs during the mentioned methodology can be overcome by utilising technical components for mixing and growth as well as transport and temperature adjustments. For the production of CdSe based semiconductor nanoparticles this method has been investigated and tuned to production amounts of kg per month. Since the use of technical components allows for easy interchange in regards of maximum through-put and size, it can be further enhanced to tens or even 100's of kgs.

Recently a consortium of U.S. and Dutch companies reported a "milestone" in high volume quantum dot manufacturing by applying the traditional high temperature dual injection method to a flow system. However as of 2011, applications using bulk-manufactured quantum dots are scarcely available.

Cadmium-Free Quantum Dots

Cadmium-free quantum dots are also called "CFQD". In many regions of the world there is now a restriction or ban on the use of heavy metals in many household goods which means that most cadmium based quantum dots are unusable for consumer-goods applications.

For commercial viability, a range of restricted, heavy metal-free quantum dots has been developed showing bright emissions in the visible and near infrared region of the spectrum and have similar optical properties to those of CdSe quantum dots. Among these systems are InP/ZnS and CuInS/ZnS, for example.

Cadmium and other restricted heavy metals used in conventional quantum dots is of a major concern in commercial applications. For Quantum Dots to be commercially viable in many applications they must not contain cadmium or other restricted metal elements.

Environmental Impact

The environmental impact of bulk manufacturing and consumption of quantum dots is currently undergoing studies in both private and public labs.

Optical Properties

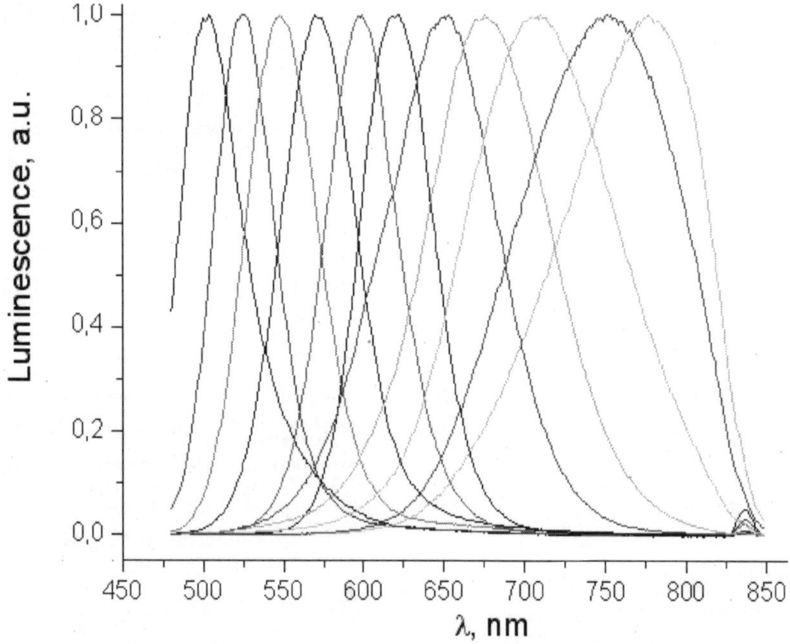

Figure: *Fluorescence spectra of CdTe quantum dots of various sizes. Different sized quantum dots emit different colour light due to quantum confinement.*

An immediate optical feature of colloidal quantum dots is their colour. While the material which makes up a quantum dot defines its intrinsic energy signature, the nanocrystal's quantum confined size is more significant at energies near the band gap. Thus quantum dots of the same material, but with different sizes, can emit light of different colours. The physical reason is the quantum confinement effect.

The larger the dot, the redder (lower energy) its fluorescence spectrum. Conversely, smaller dots emit bluer (higher energy) light. The colouration is directly related to the energy levels of the quantum dot. Quantitatively speaking, the bandgap energy that determines the energy (and hence colour) of the fluorescent light is inversely proportional to the size of the quantum dot.

Larger quantum dots have more energy levels which are also more closely spaced. This allows the quantum dot to absorb photons containing less energy, i.e., those closer to the red end of the spectrum. Recent articles in *Nanotechnology* and in other journals have begun to suggest that the shape of the quantum dot may be a factor in the colouration as well, but as yet not enough information is available.

Furthermore, it was shown that the lifetime of fluorescence is determined by the size of the quantum dot. Larger dots have more closely spaced energy levels in which the electron-hole pair can be trapped. Therefore, electron-hole pairs in larger dots live longer causing larger dots to show a longer lifetime.

As with any crystalline semiconductor, a quantum dot's electronic wave functions extend over the crystal lattice. Similar to a molecule, a quantum dot has both a quantized energy spectrum and a quantized density of electronic states near the edge of the band gap.

Quantum dots can be synthesized with larger (thicker) shells (CdSe quantum dots with CdS shells). The shell thickness has shown direct correlation to the spectroscopic properties of the particles like lifetime and emission intensity, but also to the stability.

Applications

Quantum dots are particularly significant for optical applications due to their high extinction coefficient. In electronic applications they have been proven to operate like a single electron transistor and show the Coulomb blockade effect. Quantum dots have also been suggested as implementations of qubits for quantum information processing.

The ability to tune the size of quantum dots is advantageous for many applications. For instance, larger quantum dots have a greater spectrum-shift towards red compared to smaller dots, and exhibit less pronounced quantum properties. Conversely, the smaller particles allow one to take advantage of more subtle quantum effects.

Being zero-dimensional, quantum dots have a sharper density of states than higher-dimensional structures. As a result, they have superior transport and optical properties, and are being researched for use in diode lasers, amplifiers, and biological sensors.

Quantum dots may be excited within a locally enhanced electromagnetic field produced by gold nanoparticles, which can then be observed from the surface plasmon resonance in the photoluminescent excitation spectrum of (CdSe)ZnS nanocrystals. High-quality quantum dots are well suited for optical encoding and multiplexing applications due to their broad excitation profiles and narrow/symmetric emission spectra. The new generations of quantum dots have far-reaching potential for the study of intracellular processes at the single-molecule level, high-resolution cellular imaging, long-term in vivo observation of cell trafficking, tumor targeting, and diagnostics.

Computing

Quantum dot technology is one of the most promising candidates for use in solid-state quantum computation. By applying small voltages to the leads, the flow of electrons through the quantum dot can be controlled and thereby precise measurements of the spin and other properties therein can be made. With several entangled quantum dots, or qubits, plus a way of performing operations, quantum calculations and the computers that would perform them might be possible.

Biology

In modern biological analysis, various kinds of organic dyes are used. However, with each passing year, more flexibility is being required of these dyes, and the traditional dyes are often unable to meet the expectations. To this end, quantum dots have quickly filled in the role, being found to be superior to traditional organic dyes on several counts, one of the most immediately obvious being brightness (owing to the high extinction co-efficient combined with a comparable quantum yield to fluorescent dyes) as well as their stability (allowing much less photobleaching). It has been estimated that quantum dots are 20 times brighter and 100 times more stable than traditional fluorescent reporters. For single-particle tracking, the irregular blinking of quantum dots is a minor drawback.

The usage of quantum dots for highly sensitive cellular imaging has seen major advances over the past decade. The improved photostability of quantum dots, for example, allows the acquisition of many consecutive focal-plane images that can be reconstructed into a high-resolution three-dimensional image. Another application that takes advantage of the extraordinary photostability of quantum dot probes is the real-time tracking of molecules and cells over extended periods of time.

Antibodies, streptavidin, peptides, DNA, nucleic acid aptamers, or small-molecule ligands can be used to target quantum dots to specific proteins on cells. Researchers were able to observe quantum dots in lymph nodes of mice for more than 4 months.

Semiconductor quantum dots have also been employed for in vitro imaging of pre-labeled cells. The ability to image single-cell migration in real time is expected to be important to several research areas such as embryogenesis, cancer metastasis, stem cell therapeutics, and lymphocyte immunology.

Scientists have proven that quantum dots are dramatically better than existing methods for delivering a gene-silencing tool, known as siRNA, into cells.

First attempts have been made to use quantum dots for tumor targeting under in vivo conditions. There exist two basic targeting schemes: active targeting and passive targeting. In the case of active targeting, quantum dots are functionalized with tumor-specific binding sites to selectively bind to tumor cells.

Passive targeting uses the enhanced permeation and retention of tumor cells for the delivery of quantum dot probes. Fast-growing tumor cells typically have more permeable membranes than healthy cells, allowing the leakage of small nanoparticles into the cell body. Moreover, tumor cells lack an effective lymphatic drainage system, which leads to subsequent nanoparticle-accumulation.

One of the remaining issues with quantum dot probes is their potential in vivo toxicity. For example, CdSe nanocrystals are highly toxic to cultured cells under UV illumination. The energy of UV irradiation is close to that of the covalent chemical bond energy of CdSe nanocrystals. As a result, semiconductor particles can be dissolved, in a process known as photolysis, to release toxic cadmium ions into the culture medium. In the absence of UV irradiation, however, quantum dots with a stable polymer coating have been found to be essentially nontoxic. Hydrogel encapsulation of quantum dots allows for quantum dots to be introduced into a stable aqueous solution, reducing the possibility of cadmium leakage. Then again, only little is known about the excretion process of quantum dots from living organisms. These and other questions must be carefully examined before quantum dot applications in tumor or vascular imaging can be approved for human clinical use.

Another potential cutting-edge application of quantum dots is being researched, with quantum dots acting as the inorganic fluorophore for intra-operative detection of tumors using fluorescence spectroscopy.

Photovoltaic Devices

Quantum dots may be able to increase the efficiency and reduce the cost of today's typical silicon photovoltaic cells. According to an experimental proof from 2004, quantum dots of lead selenide can produce more than one exciton from one high energy photon via the process of carrier multiplication or multiple exciton generation (MEG). This compares favourably to today's photovoltaic cells which can only manage one exciton per high-energy photon, with high kinetic energy carriers losing their energy as heat. Quantum dot photovoltaics would theoretically be cheaper to manufacture, as they can be made "using simple chemical reactions."

Light Emitting Devices

There are several inquiries into using quantum dots as light-emitting diodes to make displays and other light sources, such as "QD-LED" displays, and "QD-WLED" (White LED). In June 2006, QD Vision announced technical success in making a proof-of-concept quantum dot display and show a bright emission in the visible and near infra-red region of the spectrum.

Quantum dots are valued for displays, because they emit light in very specific gaussian distributions. This can result in a display that more accurately renders the colours that the human eye can perceive. Quantum dots also require very little power since they are not colour filtered. Additionally, since the discovery of "white-light emitting" QD, general solid-state lighting applications appear closer than ever.

A colour liquid crystal display (LCD), for example, is usually backlit by fluorescent lamps (CCFLs) or conventional white LEDs that are colour filtered to produce red, green, and blue pixels. A better solution is using a conventional blue-emitting LED as light source and converting part of the emitted light into *pure* green and red light by the appropriate quantum dots placed in front of the blue LED. This type of white light as backlight of an LCD panel allows for the best colour gamut at lower cost than a RGB LED combination using three LEDs.

Quantum dot displays that intrinsically produce monochromatic light can be more efficient, since more of the light produced reaches the eye. QD-LEDs can be fabricated on a silicon substrate, which allows integration of light sources onto silicon-based integrated circuits or microelectromechanical systems. A QD-LED integrated at a scanning microscopy tip was used to demonstrate fluorescence near-field scanning optical microscopy (NSOM) imaging.

Photodetector Devices

Quantum dot photodetectors (QDPs) can be fabricated either via solution-processing, or from conventional single-crystalline semiconductors. Conventional single-crystalline semiconductor QDPs are precluded from integration with flexible organic electronics due to the incompatibility of their growth conditions with the process windows required by organic semiconductors. On the other hand, solution-processed QDPs can be readily integrated with an almost infinite variety of substrates, and also postprocessed atop other integrated circuits. Such colloidal QDPs have potential applications in surveillance, machine vision, industrial inspection, spectroscopy, and fluorescent biomedical imaging.

Theoretical Models

A variety of theoretical frameworks exist to model optical, electronic, and structural properties of quantum dots. These may be broadly divided into quantum mechanical, semiclassical, and classical.

Quantum Mechanics

Quantum mechanical models and simulations of quantum dots often involve the interaction of electrons with a pseudopotential.

Semiclassical

Semiclassical models of quantum dots frequently incorporate a chemical potential. For example, The thermodynamic chemical potential of an N-particle system is given by

$$\mu(N) = E(N) - E(N-1)$$

whose energy terms may be obtained as solutions of the Schrodinger equation. The definition of capacitance,

$$\frac{1}{C} \equiv \frac{\Delta V}{\Delta Q},$$

with the potential difference

$$\Delta V = \frac{\Delta\mu}{e} = \frac{\mu(N+\Delta N) - \mu(N)}{e}$$

may be applied to a quantum dot with the addition or removal of individual electrons,

$$\Delta N = 1 \text{ and } \Delta Q = e.$$

Then

$$C(N) = \frac{e^2}{\mu(N+1) - \mu(N)} = \frac{e^2}{E(N)}$$

is the "quantum capacitance" of a quantum dot.

Classical Mechanics

Classical models of electrostatic properties of electrons in quantum dots are similar in nature to the Thomson problem of optimally distributing electrons on a unit sphere.

The classical electrostatic treatment of electrons confined to spherical quantum dots is similar to their treatment in the Thomson, or plum pudding model, of the atom.

The classical treatment of both two-dimensional and three-dimensional quantum dots exhibit electron shell-filling behaviour. A "periodic table of classical artificial atoms" has been described for two-dimensional quantum dots. As well, several connections have been reported between the three-dimensional Thomson problem and electron shell-filling patterns found in naturally-occurring atoms found throughout the periodic table. This latter work originated in classical electrostatic modelling of electrons in a spherical quantum dot represented by an ideal dielectric sphere.

8

Optical Fiber Sensors Based on Nanostructured Coatings

A fiber optic sensor is a sensor that uses optical fiber either as the sensing element ("intrinsic sensors"), or as a means of relaying signals from a remote sensor to the electronics that process the signals ("extrinsic sensors"). Fibers have many uses in remote sensing. Depending on the application, fiber may be used because of its small size, or because no electrical power is needed at the remote location, or because many sensors can be multiplexed along the length of a fiber by using light wavelength shift for each sensor, or by sensing the time delay as light passes along the fiber through each sensor. Time delay can be determined using a device such as an optical time-domain reflectometre and wavelength shift can be calculated using an instrument implementing optical frequency domain reflectometry.

Fiber optic sensors are also immune to electromagnetic interference, and do not conduct electricity so they can be used in places where there is high voltage electricity or inflammable material such as jet fuel. Fiber optic sensors can be designed to withstand high temperatures as well.

Intrinsic Sensors

Optical fibers can be used as sensors to measure strain, temperature, pressure and other quantities by modifying a fiber so that the quantity to be measured modulates the intensity, phase, polarization, wavelength or transit time of light in the fiber. Sensors that vary the intensity of light are the simplest, since only a simple source and detector are required. A particularly useful feature of intrinsic fiber optic sensors is that they can, if required, provide distributed sensing over very large distances.

Temperature can be measured by using a fiber that has evanescent loss that varies with temperature, or by analyzing the Raman scattering of the optical fiber. Electrical voltage can be sensed by nonlinear optical effects in specially-doped fiber, which alter the polarization of light as a function of voltage or electric field. Angle measurement sensors can be based on the Sagnac effect.

Special fibers like long-period fiber grating (LPG) optical fibers can be used for direction recognition . Photonics Research Group of Aston University in UK has some publications on vectorial bend sensor applications.

Optical fibers are used as hydrophones for seismic and sonar applications. Hydrophone systems with more than one hundred sensors per fiber cable have been developed. Hydrophone sensor systems are used by the oil industry as well as a few countries' navies. Both bottom-mounted hydrophone arrays and towed streamer systems are in use. The German company Sennheiser developed a laser microphone for use with optical fibers.

A fiber optic microphone and fiber-optic based headphone are useful in areas with strong electrical or magnetic fields, such as communication amongst the team of people working on a patient inside a magnetic resonance imaging (MRI) machine during MRI-guided surgery.

Optical fiber sensors for temperature and pressure have been developed for downhole measurement in oil wells. The fiber optic sensor is well suited for this environment as it functions at temperatures too high for semiconductor sensors (distributed temperature sensing).

Optical fibers can be made into interferometric sensors such as fiber optic gyroscopes, which are used in the Boeing 767 and in some car models (for navigation purposes). They are also used to make hydrogen sensors.

Fiber-optic sensors have been developed to measure co-located temperature and strain simultaneously with very high accuracy using fiber Bragg gratings. This is particularly useful when acquiring information from small complex structures. Brillouin scattering effects can be used to detect strain and temperature over larger distances (20–30 kilometres).

Other Examples

A fiber-optic AC/DC voltage sensor in the middle and high voltage range (100–2000 V) can be created by inducing measurable amounts of Kerr nonlinearity in single mode optical fiber by exposing a calculated length of fiber to the external electric field. The measurement technique is based on polarimetric detection and high accuracy is achieved in a hostile industrial environment.

High frequency (5 MHz–1 GHz) electromagnetic fields can be detected by induced nonlinear effects in fiber with a suitable structure. The fiber used is designed such that the Faraday and Kerr effects cause considerable phase change in the presence of the external field. With appropriate sensor design, this type of fiber can be used to measure different electrical and magnetic quantities and different internal parameters of fiber material.

Electrical power can be measured in a fiber by using a structured bulk fiber ampere sensor coupled with proper signal processing in a polarimetric detection scheme. Experiments have been carried out in support of the technique.

Fiber-optic sensors are used in electrical switchgear to transmit light from an electrical arc flash to a digital protective relay to enable fast tripping of a breaker to reduce the energy in the arc blast.

Extrinsic Sensors

Extrinsic fiber optic sensors use an optical fiber cable, normally a multimode one, to transmit modulated light from either a non-fiber optical sensor, or an electronic sensor connected to an optical transmitter. A major benefit of extrinsic sensors is their ability to reach places which are otherwise inaccessible. An example is the measurement of temperature inside aircraft jet engines by using a fiber to transmit radiation into a radiation pyrometre located outside the engine. Extrinsic sensors can also be used in the same way to measure the internal temperature of electrical transformers, where the extreme electromagnetic fields present make other measurement techniques impossible.

Extrinsic fiber optic sensors provide excellent protection of measurement signals against noise corruption. Unfortunately, many conventional sensors produce electrical output which must be converted into an optical signal for use with fiber.

For example, in the case of a platinum resistance thermometre, the temperature changes are translated into resistance changes. The PRT must therefore have an electrical power supply. The modulated voltage level at the output of the PRT can then be injected into the optical fiber via the usual type of transmitter. This complicates the measurement process and means that low-voltage power cables must be routed to the transducer.

Extrinsic sensors are used to measure vibration, rotation, displacement, velocity, acceleration, torque, and twisting.

Distributed Temperature Sensing

Distributed temperature sensing systems (DTS) are optoelectronic devices which measure temperatures by means of optical fibres functioning as linear sensors. Temperatures are recorded along the optical sensor cable, thus not at points, but as a continuous profile. A high accuracy of temperature determination is achieved over great distances. Typically the DTS systems can locate the temperature to a spatial resolution of 1 m with accuracy to within ±1°C at a resolution of 0.01°C. Measurement distances of greater than 30 km can be monitored and some specialised systems can provide even tighter spatial resolutions.

Measuring Principle—Raman Effect

Physical measurement dimensions, such as temperature or pressure and tensile forces, can affect glass fibres and locally change the characteristics of light transmission in the fibre. As a result of the damping of the light in the quartz glass fibres through scattering, the location of an external physical effect can be determined so that the optical fibre can be employed as a linear sensor. Optical fibres are made from doped quartz glass.

Quartz glass is a form of silicon dioxide (SiO_2) with amorphous solid structure. Thermal effects induce lattice oscillations within the solid. When light falls onto these thermally excited molecular oscillations, an interaction occurs between the light particles (photons) and the electrons of the molecule. Light scattering, also known as Raman scattering, occurs in the optical fibre. Unlike incident light, this scattered light undergoes a spectral shift by an amount equivalent to the resonance frequency of the lattice oscillation. The light scattered back from the fibre optic therefore contains three different spectral shares:

- the Rayleigh scattering with the wavelength of the laser source used,
- the Stokes line components from photons shifted to longer wavelength (lower frequency), and
- the anti-Stokes line components with photons shifted to shorter wavelength (higher frequency) than the Rayleigh scattering.

The intensity of the so-called anti-Stokes band is temperature-dependent, while the so-called Stokes band is practically independent of temperature. The local temperature of the optical fibre is derived from the ratio of the anti-Stokes and Stokes light intensities.

Measuring Principle—OTDR and OFDR Technology

There are two basic principles of measurement for distributed sensing technology, OTDR (Optical Time Domain Reflectometry) and OFDR (Optical

Frequency Domain Reflectometry). For Distributed Temperature Sensing often a Code Correlation technology is employed which carries elements from both principles.

OTDR was developed more than 20 years ago and has become the industry standard for telecom loss measurements which detects the—compared to Raman signal very dominant—Rayleigh backscattering signals. The principle for OTDR is quite simple and is very similar to the time of flight measurement used for radar. Essentially a narrow laser pulse generated either by semiconductor or solid state lasers is sent into the fibre and the backscattered light is analysed. From the time it takes the backscattered light to return to the detection unit it is possible to locate the location of the temperature event.

Alternative DTS evaluation units deploy the method of Optical Frequency Domain Reflectometry (OFDR). The OFDR system provides information on the local characteristic only when the backscatter signal detected during the entire measurement time is measured as a function of frequency in a complex fashion, and then subjected to Fourier transformation. The essential principles of OFDR technology are the quasi continuous wave mode employed by the laser and the narrow-band detection of the optical back scatter signal. This is offset by the technically difficult measurement of the Raman scatter light and rather complex signal processing, due to the FFT calculation with higher linearity requirements for the electronic components.

Code Correlation DTS sends on/off sequences of limited length into the fiber. The codes are chosen to have suitable properties, e.g. Binary Golay code. In contrast to OTDR technology, the optical energy is spread over a code rather than packed into a single pulse. Thus a light source with lower peak power compared to OTDR technology can be used, e.g. long life compact semiconductor lasers. The detected backscatter needs to be transformed—similar to OFDR technology—back into a spatial profile, e.g. by cross-correlation. In contrast to OFDR technology, the emission is finite (for example 128 bit) which avoids that weak scattered signals from far are superposed by strong scattered signals from short distance, improving the Shot noise and the signal-to-noise ratio.

Using theses techniques it is possible to analyse distances of greater than 30 km from one system and to measure temperature resolutions of less than 0.01°C.

Construction of Sensing Cable & System Integration

The temperature measuring system consists of a controller (laser source, pulse generator for OTDR or code generator for Code Correlation or modulator and HF mixer for OFDR, optical module, receiver and micro-processor unit) and a quartz glass fibre as line-shaped temperature sensor. The fibre optic cable (can be 30 km+ in length) is passive in nature and has no individual sensing points and therefore can be manufactured based on standard telecoms fibres. This offers excellent economies of scale.

Because the system designer/integrator does not have to worry about the precise location of each sensing point the cost for designing and installing a sensing system based on distributed fibre optic sensors is greatly reduced from that of traditional sensors. Additionally, because the sensing cable has no moving parts and design lives of 30 years +, the maintenance and operation costs are also considerably less than for conventional sensors. Additional benefits of fibre optic sensing technology are that it is immune to electromagnetic interference, vibration and is safe for use in hazardous zones (the laser power falls below the levels that can cause ignition), thus making these sensors ideal for use in industrial sensing applications.

With regards to the construction of the sensing cable, although it is based on standard fibre optics, care must be taken in the design of the individual sensing cable to ensure that adequate protection is provided for the fibre. This must take into account operating temperature (standard cables operate to 85°C but it is possible to measure up to 700°C with the correct design), gaseous environment (hydrogen can cause deterioration of the measurement though "hydrogen darkening" - aka attenuation - of the silica glass compounds) and mechanical protection.

Most of the available DTS systems have flexible system architectures and are relatively simple to integrate into industrial control systems such as SCADA. In the oil and gas industry an XML based file standard (WITSML) has been developed for transfer of data from DTS instruments. The standard is maintained by Energistics.

Laser Safety and Operation of System

When operating a system based on optical measurements such as optical DTS, laser safety requirements need to be considered for permanent installations. Many systems use low power laser design, e.g. with classification as laser safety class 1M, which can be applied by anyone (no approved laser safety officers required). Some systems are based on higher power lasers of a 3B rating, which although safe for use by approved laser safety officers, may not be suitable for permanent installations.

The advantage of purely passive optical sensor technology is the lack of electric or electromagnetic interaction. Some DTS systems on the market use a special low power design and are inherently safe in explosive environments, e.g. certified to ATEX directive Zone 0.

For use in fire detection application, regulations usually require certified systems according to relevant standards, such as EN 54-5 or EN 54-22 (Europe), UL521 or FM (USA), cUL521 (Canada) and/or other national or local standards.

Temperature Estimations by Using DTS

Temperature distributions can be used to develop models based on the Proper Orthogonal Decomposition Method or principal component analysis. This allows to reconstruct the temperature distribution by measuring only in a few spatial locations

Applications

Distributed temperature sensing can be deployed successfully in multiple industrial segments:

- Oil and gas production—permanent downhole monitoring, coil tubing optical enabled deployed intervention systems, slickline optical cable deployed intervention systems.
- Power cable and transmission line monitoring (ampacity optimisation)
- Fire detection in tunnels, industrial conveyor belts and special hazard buildings
- Industrial induction furnace surveillance
- Integrity of liquid natural gas (LNG) carriers and terminals
- Leakage detection at dikes and dams
- Temperature monitoring in plant and process engineering, including transmission pipelines
- Storage tanks and vessels

More recently, DTS has been applied for ecological monitoring as well:

- Stream temperature
- Groundwater source detection
- Temperature profiles in a mine shaft and over lakes and glaciers
- Deep rainforest ambient temperature at various foliage densities
- Temperature profiles in an underground mine, Australia
- Temperature profiles in ground loop heat exchangers (used for ground coupled heating and cooling systems)

Distributed Acoustic Sensing

Rayleigh scattering based distributed acoustic sensing (DAS) systems use fibre optic cables to provide distributed strain sensing. In DAS, the optical fiber cable becomes the sensing element and measurements are made, and in part processed, using an attached optoelectronic device. Such a system allows acoustic frequency strain signals to be detected over large distances and in harsh environments.

Fundamentals of Rayleigh Scatter Based Fibre Optic Sensing

In Rayleigh scatter based distributed fibre optic sensing, a coherent laser pulse is sent along an optic fibre, and scattering sites within the fibre cause the fibre to act as a distributed interferometre with a gauge length approximately equal to the pulse length. The intensity of the reflected light is measured as a function of time after transmission of the laser pulse. When the pulse has had time to travel the full length of the fibre and back, the next laser pulse can be sent along the fibre. Changes in the reflected intensity of successive pulses from the same region of fibre are caused by changes in the optical path length of that section of fibre. This type of system is very sensitive to both strain and temperature variations of the fibre and measurements can be made simultaneously at all sections of the fibre.

Capabilities of Rayleigh-Based Systems

Maximum Range: The optical pulse is attenuated as it propagates it along the fibre. For a single mode fibre operating at 1550 nm, a typical attenuation is 0.2 dB/km. Since the light must make a double pass along each section of fibre, an extra 1 km of it induces an additional loss of 0.4dB. The maximum range of the system occurs when the amplitude of the reflected pulse becomes so low it is impossible to obtain a clear signal from it. It is not possible to counteract this effect by increasing the input power because above a certain level as this will induce nonlinear optical effects which will disrupt the operation of the system. Typically the maximum range that can be measured is around 40–50 km.

Strain Resolution

The minimum value of strain that can be measured depends on the carrier to noise ratio of the returning optical signal. The carrier level is largely determined by the amplitude of the optical signal while the noise is a combination of that from a variety of sources including laser noise, electronic noise and detector noise.

Spatial Resolution and Spatial Sampling Period

The spatial resolution is mainly determined by the duration of the transmitted pulse, with a 100ns pulse giving 10m resolution being a typical value. The amount of reflected light is proportional to the pulse length so there is a trade-off between spatial resolution and maximum range. To improve the maximum range it would be desirable to use a longer pulse length to increase the reflected light level but this leads to a larger spatial resolution.

In order for two signals to be independent they must be obtained from two points on the fibre that are separated by at least the spatial resolution. It is possible to obtain samples at separations less than the spatial resolution and although this produces signals that are not independent of each other such an approach does offer advantages in some applications. The separation between the sampling points is sometimes referred to as the spatial sampling period.

Acquisition Rate

Before the next laser pulse can be transmitted the previous one must have had time to travel to the far end of the fibre and for the reflections from there to return, otherwise reflections would be returning from different sections of the fibre at the same time and the system would not operate properly. For a fibre 50 km long the maximum pulse rate is just over 2 kHz. Therefore strains can be measured which vary at frequencies up to the Nyquist frequency of 1 kHz. Shorter fibres clearly enable higher acquisition rates.

Temperature Measurements

Although the system is sensitive to both temperature and strain variations these can often be separated as those due to temperature tend to occur at a lower frequency range than strain. Unlike other distributed fibre techniques such as those based on Brillouin or Raman scatter, distributed acoustic sensing is only able to detect changes in temperature rather than its absolute value.

Comparison with Other Fibre Optic Distributed Sensing Techniques

Distributed Acoustic Sensing relies on light which is Rayleigh backscattered from small variations in the refractive index of the fibre. The backscattered light has the same frequency as the transmitted light. There are a number of other distributed fibre sensing techniques that rely on different scattering mechanisms and can be used to measure other parameters.

Brillouin scatter occurs due to the interaction between the light and acoustic phonons travelling in the fibre. As the light is scattered by a moving phonon its frequency is shifted by the Doppler effect by around 10 GHz. Light is generated at both at above (anti-Stokes shift) and below (Stokes shift) the original optical frequency. The intensity and frequency shifts of the two components are dependent on both temperature and strain and by measuring the shifts, absolute values of the two parameters can be calculated using a

Distributed Temperature and Strain Sensing (DTSS) system. Brillouin scatter is much weaker than Rayleigh scatter and so the reflections from a number of pulses must be summed together to enable the measurements to be made. Therefore the maximum frequency at which changes can be measured using Brillouin scatter is typically a few 10's of Hz. Raman scatter occurs when light is scattered in interaction with molecular vibrations in the fiber.

As with Brillouin scattering both Stokes and anti-Stokes components are produced and these are shifted from the wavelength of the incident light by several tens of nanometres. By measuring the ratio in intensity between the Stokes and anti-Stokes components an absolute value of temperature can be measured by a distributed temperature sensing (DTS) system.

The larger wavelength shifts compared to Brillouin scatter mean that it is easier to separate the scattered Raman light from the un-shifted Rayleigh scattered component. However the intensity of the Raman scatter is even lower than the Brillouin scatter and so it is normally necessary to average for many seconds or even minutes in order to get reasonable results. Therefore Raman based systems are only suitable for measuring slowly varying temperatures.

Applications

The sensitivity and speed of Rayleigh based sensing allows distributed acoustic monitoring over distances of up to 100 km from each laser source. With suitable analysis software, continuous monitoring of pipelines for unwanted interference, as well as leaks or flow irregularities is possible.

Roads, borders, perimeters etc. can be monitored for unusual activity with the position of the activity being determined to within approximately 10 metres. Due to the ability of the optic fibre to operate in harsh environments, the technology can also be used in oil well monitoring applications, allowing real-time information on the state of the well to be determined.

Strain Gauge

A strain gauge is a device used to measure strain on an object. Invented by Edward E. Simmons and Arthur C. Ruge in 1938, the most common type of strain gauge consists of an insulating flexible backing which supports a metallic foil pattern.

The gauge is attached to the object by a suitable adhesive, such as cyanoacrylate. As the object is deformed, the foil is deformed, causing its electrical resistance to change. This resistance change, usually measured using a Wheatstone bridge, is related to the strain by the quantity known as the *gauge factor*.

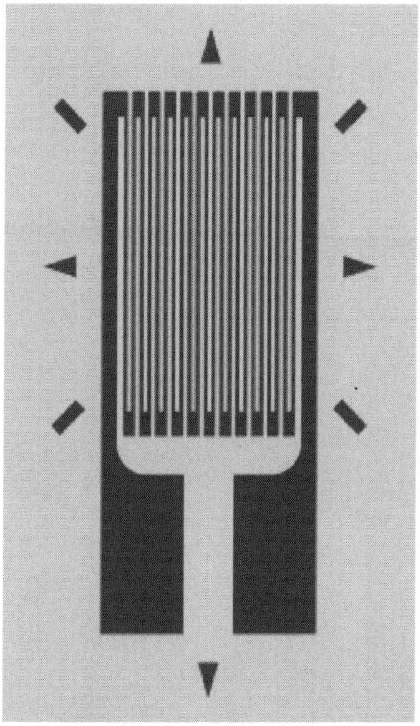

Figure: *Typical foil strain gauge. The gauge is far more sensitive to strain in the vertical direction than in the horizontal direction. The markings outside the active area help to align the gauge during installation.*

Physical Operation

A strain gauge takes advantage of the physical property of electrical conductance and its dependence on the conductor's geometry. When an electrical conductor is stretched within the limits of its elasticity such that it does not break or permanently deform, it will become narrower and longer, changes that increase its electrical resistance end-to-end. Conversely, when a conductor is compressed such that it does not buckle, it will broaden and shorten, changes that decrease its electrical resistance end-to-end. From the measured electrical resistance of the strain gauge, the amount of applied stress may be inferred. A typical strain gauge arranges a long, thin conductive strip in a zig-zag pattern of parallel lines such that a small amount of stress in the direction of the orientation of the parallel lines results in a multiplicatively larger strain measurement over the effective length of the conductor surfaces in the array of conductive lines—and hence a multiplicatively larger change in resistance—than would be observed with a single straight-line conductive wire.

Gauge Factor

The gauge factor GF is defined as:

$$GF = \frac{\Delta R / R_G}{\epsilon}$$

where

ΔR is the change in resistance caused by strain,

R_G is the resistance of the undeformed gauge, and

ϵ is strain.

For metallic foil gauges, the gauge factor is usually a little over 2. For a single active gauge and three dummy resistors, the output v from the bridge is:

$$v = \frac{BV \cdot GF \cdot \epsilon}{4}$$

where

BV is the bridge excitation voltage.

Foil gauges typically have active areas of about 2–10 mm^2 in size. With careful installation, the correct gauge, and the correct adhesive, strains up to at least 10% can be measured.

Gauges in Practice

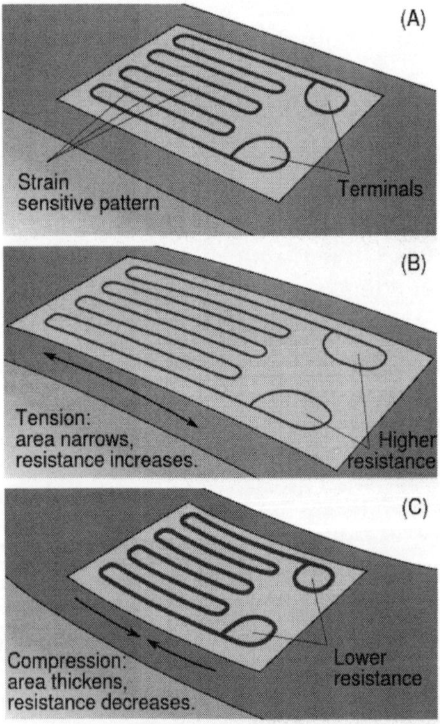

Figure: *Visualization of the working concept behind the strain gauge on a beam under exaggerated bending.*

An excitation voltage is applied to input leads of the gauge network, and a voltage reading is taken from the output leads. Typical input voltages are 5 V or 12 V and typical output readings are in millivolts.

Foil strain gauges are used in many situations. Different applications place different requirements on the gauge. In most cases the orientation of the strain gauge is significant.

Gauges attached to a load cell would normally be expected to remain stable over a period of years, if not decades; while those used to measure response in a dynamic experiment may only need to remain attached to the object for a few days, be energized for less than an hour, and operate for less than a second.

Strain gauges are attached to the substrate with a special glue. The type of glue depends on the required lifetime of the measurement system. For short term measurements (up to some weeks) cyanoacrylic glue is appropriate, for long lasting installation epoxy glue is required. Usually epoxy glue requires high temperature curing (at about 80-100°C). The preparation of the surface where the strain gauge is to be glued is of the utmost importance. The surface must be smoothed (e.g. with very fine sand paper), deoiled with solvents, the solvent traces must then be removed and the strain gauge must be glued immediately after this to avoid oxidation or pollution of the prepared area. If these steps are not followed the strain gauge binding to the surface may be unreliable and unpredictable measurement errors may be generated.

Strain gauge based technology is utilised commonly in the manufacture of pressure sensors. The gauges used in pressure sensors themselves are commonly made from silicon, polysilicon, metal film, thick film, and bonded foil.

Variations in Temperature

Variations in temperature will cause a multitude of effects. The object will change in size by thermal expansion, which will be detected as a strain by the gauge. Resistance of the gauge will change, and resistance of the connecting wires will change.

Most strain gauges are made from a constantan alloy. Various constantan alloys and Karma alloys have been designed so that the temperature effects on the resistance of the strain gauge itself cancel out the resistance change of the gauge due to the thermal expansion of the object under test. Because different materials have different amounts of thermal expansion, self-temperature compensation (STC) requires selecting a particular alloy matched to the material of the object under test.

Strain gauges that are not self-temperature-compensated (such as isoelastic alloy) can be temperature compensated by use of the dummy gauge technique. A dummy gauge (identical to the active strain gauge) is installed on an unstrained sample of the same material as the test specimen. The sample with the dummy gauge is placed in thermal contact with the test specimen, adjacent to the active gauge. The dummy gauge is wired into a Wheatstone bridge on an adjacent arm to the active gauge so that the temperature effects on the active and dummy gauges cancel each other. (Murphy's Law was originally coined in response to a set of gauges being incorrectly wired into a Wheatstone bridge.)

Temperature effects on the lead wires can be cancelled by using a "3-wire bridge" or a –4-wire ohm circuit" (also called a –4-wire Kelvin connection").

In any case it is a good engineering practice to keep the Wheatstone bridge voltage drive low enough to avoid the self heating of the strain gauge. The self heating of the strain gauge depends on its mechanical characteristic (large strain gauges are less prone to self heating). Low voltage drive levels of the bridge reduce the sensitivity of the overall system.

Errors and Compensation

Zero Offset: If the impedance of the four gauge arms are not exactly the same after bonding the gauge to the force collector, there will be a zero offset which can be compensated by introducing a parallel resistor to one or more of the gauge arms.

Temperature Coefficient of Gauge Factor (TCGF): This is the change of sensitivity of the device to strain with change in temperature. This is generally compensated for by the introduction of a fixed resistance in the input leg, whereby the effective supplied voltage will increase with temperature, compensating for the decrease in sensitivity with temperature.

Zero Shift with Temperature: If the TCGF of each gauge is not the same, there will be a zero shift with temperature. This is also caused by anomalies in the force collector. This is usually compensated for with one or more resistors strategically placed in the compensation network.

Linearity: This is an error whereby the sensitivity changes across the pressure range. This is commonly a function of the force collection thickness selection for the intended pressure and/or the quality of the bonding.

Hysteresis: This is an error of return to zero after pressure excursion.

Repeatability: This error is sometimes tied-in with hysteresis but is across the pressure range.

EMI Induced Errors: As strain gauges output voltage is in the mV range, even μV if the Wheatstone bridge voltage drive is kept low to avoid self heating of the element, special care must be taken in output signal amplification to avoid amplifying also the superimposed noise. A solution which is frequently adopted is to use "carrier frequency" amplifiers which convert the voltage variation into a frequency variation (as in VCOs) and have a narrow bandwidth thus reducing out of band EMI.

Overloading: If a strain gauge is loaded beyond its design limit (measured in microstrain) its performance degrades and can not be recovered. Normally good engineering practice suggests not to stress strain gauges beyond +/-3000 microstrain.

Humidity: If the wires connecting the strain gauge to the signal conditioner are not protected against humidity (bare wire) a parasitic resistance creates between the wires and the substrate to which the strain gauge is glued, or between the two wires themselves. This resistance introduces an error which is proportional to the resistance of the strain gauge. For this reason low resistance strain gauges (120 ohm) are less prone to this type of error. To avoid this error it is sufficient to protect the strain gauges wires with insulating enamel (e.g., epoxy or polyurethanic type). Strain gauges with unprotected wires may be used only in a dry laboratory environment but not in an industrial one.

In some applications, strain gauges add mass and damping to the vibration profiles of the hardware they are intended to measure. In the turbomachinery industry, one used alternative to strain gauge technology in the measurement of vibrations on rotating hardware is the Non-Intrusive Stress Measurement System, which allows measurement of blade vibrations without any blade or disc-mounted hardware.

Other Gauge Types

For measurements of small strain, semiconductor strain gauges, so called piezoresistors, are often preferred over foil gauges. A semiconductor gauge usually has a larger gauge factor than a foil gauge. Semiconductor gauges tend to be more expensive, more sensitive to temperature changes, and are more fragile than foil gauges.

In biological measurements, especially blood flow / tissue swelling, a variant called mercury-in-rubber strain gauge is used. This kind of strain gauge consists of a small amount of liquid mercury enclosed in a small rubber tube, which is applied around e.g., a toe or leg. Swelling of the body part results in stretching of the tube, making it both longer and thinner, which increases electrical resistance.

Fiber optic sensing can be employed to measure strain along an optical fiber. Measurements can be distributed along the fiber, or taken at predetermined points on the fiber. The 2010 America's Cup boats Alinghi 5 and USA-17 both employ embedded sensors of this type.

Capacitive strain gauges use a variable capacitor to indicate the level of mechanical deformation.

Vibrating wire strain gauges are used in Geotechnical and Civil Engineering applications. The gauge consists of a vibrating, tensioned wire. The strain is calculated by measuring the resonant frequency of the wire (an increase in tension increases the resonant frequency).

Mechanical Types

Simple mechanical types (such as illustrated to the left) are used in civil engineering to measure movement of buildings, foundations, and other structures. In the illustrated example, the two halves of the device are rigidly attached to the foundation wall on opposite sides of the crack. The red reference lines are on the transparent half and the grid is on the opaque white half. Both vertical and horizontal movement can be monitored over time. In this picture, the crack can be seen to have widened by approximately 0.3 mm (with no vertical movement) since the gauge was installed.

More sophisticated mechanical types incorporate dial indicators and mechanisms to compensate for temperature changes. These types can measure movements as small as 0.002 mm.

Piezoresistive Effect

The piezoresistive effect describes change in the electrical resistivity of a semiconductor or metal when mechanical strain is applied. In contrast to the piezoelectric effect, the piezoresistive effect only causes a change in electrical resistance, not in electric potential.

The change of electrical resistance in metal devices due to an applied mechanical load was first discovered in 1856 by Lord Kelvin. With single crystal silicon becoming the material of choice for the design of analog and digital circuits, the large piezoresistive effect in silicon and germanium was first discovered in 1954 (Smith 1954).

Mechanism

In conducting and semi-conducting materials, changes in inter-atomic spacing resulting from strain affect the bandgaps, making it easier (or harder depending on the material and strain) for electrons to be raised into the conduction band. This results in a change in resistivity of the material. Within a certain range of strain this relationship is linear, so that the piezoresistive coefficient

$$\rho_\sigma = \frac{\left(\dfrac{\partial\rho}{\rho}\right)}{\varepsilon}$$

where

$\partial\rho$ = Change in resistivity

ρ = Original resistivity

ε = Strain

is constant.

Piezoresistivity in Metals

Usually the resistance change in metals is mostly due to the change of geometry resulting from applied mechanical stress. However, even though the piezoresistive effect is small in those cases it is often not negligible. In cases where it is, it can be calculated using the simple resistance equation derived from ohm's law;

$$R = \rho \frac{\ell}{A}$$

where

ℓ Conductor length [m]

A Cross-sectional area of the current flow [m^2]

Some metals display piezoresistivity that is much larger than the resistance change due to geometry. In Platinum alloys, for instance, piezoresistivity is more than a factor of two larger, combining with the geometry effects to give a strain gauge sensitivity of up to more than three times as large than due to geometry effects alone. Pure Nickel's piezoresistivity is -13 times larger, completely dwarfing and even reversing the sign of the geometry induced resistance change.

Piezoresistive Effect in Semiconductors

The piezoresistive effect of semiconductor materials can be several orders of magnitudes larger than the geometrical effect and is present in materials like germanium, polycrystalline silicon, amorphous silicon, silicon carbide, and single crystal silicon. Hence, semiconductor strain gauges with a very high coefficient of sensitivity can be built. For precision measurements they are more difficult to handle than metal strain gauges, because semiconductor strain gauges are generally more sensitive to environmental conditions (esp. temperature).

For silicon, gauge factors can be two orders of magnitudes larger than those observed in most metals (Smith 1954). The resistance of n-conducting silicon mainly changes due to a shift of the three different conducting valley pairs. The shifting causes a redistribution of the carriers between valleys with different mobilities. This results in varying mobilities dependent on the direction of current flow. A minor effect is due to the effective mass change related to changing shapes of the valleys. In p-conducting silicon the phenomena are more complex and also result in mass changes and hole transfer.

Piezoresistive Silicon Devices

The piezoresistive effect of semiconductors has been used for sensor devices employing all kinds of semiconductor materials such as germanium, polycrystalline silicon, amorphous silicon, and single crystal silicon. Since silicon is today the material of choice for integrated digital and analog circuits the use of piezoresistive silicon devices has been of great interest. It enables the easy integration of stress sensors with Bipolar and CMOS circuits.

This has enabled a wide range of products using the piezoresistive effect. Many commercial devices such as pressure sensors and acceleration sensors employ the piezoresistive effect in silicon.

But due to its magnitude the piezoresistive effect in silicon has also attracted the attention of research and development for all other devices using single crystal silicon. Semiconductor Hall sensors, for example, were capable of achieving their current precision only after employing methods which eliminate signal contributions due the applied mechanical stress.

Piezoresistors

Piezoresistors are resistors made from a piezoresistive material and are usually used for measurement of mechanical stress. They are the simplest form of piezoresistive devices.

Fabrication

Piezoresistors can be fabricated using wide variety of piezoresistive materials. The simplest form of piezoresistive silicon sensors are diffused resistors. Piezoresistors consist of a simple two contact diffused n- or p-wells within a p- or n-substrate. As the typical square resistances of these devices are in the range of several hundred ohms, additional p+ or n+ plus diffusions are necessary to facilitate ohmic contacts to the device.

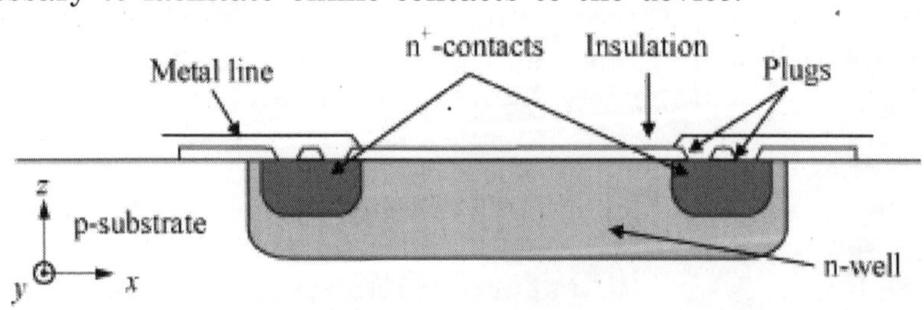

Figure: *Schematic cross-section of the basic elements of a silicon n-well piezoresistor.*

Physics of Operation

For typical stress values in the MPa range the stress dependent voltage drop along the resistor Vr, can be considered to be linear. A piezoresistor aligned with the x-axis as shown in the figure may be described by

$$V_r = R_0 I[1 + \pi_L \sigma_{xx} + \pi_T (\sigma_{yy} + \sigma_{zz})]$$

where R_0, I, π_T, π_L, and σ_{ij} denote the stress free resistance, the applied current, the transverse and longitudinal piezoresistive coefficients, and the three tensile stress components, respectively. The piezoresistive coefficients vary significantly with the sensor orientation with respect to the crystallographic axes and with the doping profile. Despite the fairly large stress sensitivity of simple resistors, they are preferably used in more complex configurations eliminating certain cross sensitivities and drawbacks. Piezoresistors have the disadvantage of being highly sensitive to temperature changes while featuring comparatively small relative stress dependent signal amplitude changes.

Other Piezoresistive Devices

In silicon the piezoresistive effect is used in piezoresistors, transducers, piezo-FETS, solid state accelerometres and bipolar transistors.

Nanocircuitry

Nanocircuits are electrical circuits operating on the nanometre scale. This is well into the quantum realm, where quantum mechanical effects become very important. One nanometre is equal to 10^{-9} metres or a row of 10 hydrogen atoms. With such progressively smaller circuits, more can be fitted on a computer chip. This allows faster and more complex functions using less power. Nanocircuits are composed of three different fundamental components. These are transistors, interconnections, and architecture, all fabricated on the nanometre scale.

Various Approaches to Nanocircuitry

A variety of proposals have been made to implement nanocircuitry in different forms. These include Single-Electron Transistors, Quantum dot cellular automata, and Nanoscale Crossbar Latches. However, likely nearer-term approaches will involve incorporation of nanomaterials to improve MOSFETs. These currently form the basis of most analog and digital circuit designs, the scaling of which drives Moore's Law. A review article covering the MOSFET design and its future was published in 2004 comparing different geometries of MOSFETs under scale reduction and noted that circular cross-section vertical channel FETs are optimal for scale reduction. This configuration is capable of being implemented with a high density using vertical semiconductor cylindrical channels with nanoscale diameters and Infineon Technologies and Samsung have begun research and development in this direction resulting in some basic patents using nanowires and carbon nanotubes in MOSFET designs.

In an alternative approach, Nanosys is a new company using solution based deposition and alignment processes to pattern pre-fabricated arrays of nanowires on a substrate to serve as a lateral channel of an FET. While not capable of the same scalability as single nanowire FETs, the use of pre-fabricated multiple nanowires for the channel increases reliability and reduces production costs since large volume printing processes may be used to deposit the nanowires at a lower temperature than conventional fabrication procedures.

In addition, due to the lower temperature deposition a wider variety of materials such as polymers may be used as the carrier substrate for the transistors opening the door to flexible electronic applications such as electronic paper, bendable flat panel displays, and wide area solar cells.

Production Methods

One of the most fundamental concepts to understanding nanocircuits is the formulation of Moore's Law. This concept arose when Intel co-founder Gordon Moore became interested in the cost of transistors and trying to fit more onto one chip.

It relates that the number of transistors that can be fabricated on a silicon integrated circuit—and therefore the computing abilities of such a circuit—is doubling every 18 to 24 months. The more transistors one can fit on a circuit, the more computational abilities the computer will have. This is why scientists and engineers are working together to produce these nanocircuits so millions and perhaps even billions of transistors will be able to fit onto a chip. Despite how good this may sound, there are many problems that arise when so many transistors are packed together.

With circuits being so tiny, they tend to have more problems than larger circuits, more particularly heat - the amount of power applied over a smaller surface area makes heat dissipation difficult, this excess heat will cause errors and can destroy the chip. Nanoscale circuits are more sensitive to temperature changes, cosmic rays and electromagnetic interference than today's circuits. As more transistors are packed onto a chip, phenomena such as stray signals on the chip, the need to dissipate the heat from so many closely packed devices, tunneling across insulation barriers due to the small scale, and fabrication difficulties will halt or severely slow progress. Many believe the market for nanocircuits will reach equilibrium around 2015. At this time they believe the cost of a fabrication facility may be as much as $200 billion. There will be a time when the cost of making circuits even smaller will be too much, and the speed of computers will reach a maximum. For this reason, many scientists believe that Moore's Law will not hold forever and will soon reach a peak, since Moore's law is largely predicated on computational gains caused by improvements in micro-lithographic etching technologies.

In producing these nanocircuits, there are many aspects involved. The first part of their organisation begins with transistors. As of right now, most electronics are using silicon-based transistors. Transistors are an integral part of circuits as they control the flow of electricity and transform weak electrical signals to strong ones. They also control electric current as they can turn it on off, or even amplify signals. Circuits now use silicon as a transistor because it can easily be switched between conducting and nonconducting states. However, in nanoelectronics, transistors might be organic molecules or nanoscale inorganic structures. Semiconductors, which are part of transistors, are also being made of organic molecules in the nano state.

The second aspect of nanocircuit organisation is interconnection. This involves logical and mathematical operations and the wires linking the transistors together that make this possible. In nanocircuits, nanotubes and other wires as narrow as one nanometre are used to link transistors together. Nanowires have been made from carbon nanotubes for a few years. Until a few years ago, transistors and nanowires were put together to produce the circuit. However, scientists have been able to produce a nanowire with transistors in it. In 2004, Harvard University nanotech pioneer Charles Lieber and his team have made a nanowire—10,000 times thinner than a sheet of paper—that contains a string of transistors. Essentially, transistors and nanowires are already pre-wired so as to eliminate the difficult task of trying to connect transistors together with nanowires.

The last part of nanocircuit organisation is architecture. This has been explained as the overall way the transistors are interconnected, so that the circuit can plug into a computer or other system and operate independently of the lower-level details. With nanocircuits being so small, they are destined for error and defects. Scientists have devised a way to get around this. Their architecture combines circuits that have redundant logic gates and interconnections with the ability to reconfigure structures at several levels on a chip. The redundancy lets the circuit identify problems and reconfigure itself so the circuit can avoid more problems. It also allows for errors within the logic gate and still have it work properly without giving a wrong result.

Potential Applications and Breakthroughs

Scientists in India have recently developed the world's smallest transistor which will be used for nanocircuits. The transistor is made entirely from carbon nanotubes. Nanotubes are rolled up sheets of carbon atoms and are more than a thousand times thinner than human hair.

Normally circuits use silicon-based transistors, but these will soon replace those. The transistor has two different branches that meet at a single point, hence giving it a Y shape. Current can flow throughout both branches and is controlled by a third branch that turns the voltage on or off. This new breakthrough can now allow for nanocircuits to hold completely to their name as they can be made entirely from nanotubes. Before this discovery, logic circuits used nanotubes, but needed metal gates to be able to control the flow of electrical current.

Arguably the biggest potential application of nanocircuits deals with computers and electronics. Scientists and engineers are always looking to make computers faster. Some think in the nearer term, we could see hybrids of micro- and nano-: silicon with a nano core—perhaps a high-density computer memory that retains its contents forever.

Unlike conventional circuit design, which proceeds from blueprint to photographic pattern to chip, nanocircuit design will probably begin with the chip—a haphazard jumble of as many as 1024 components and wires, not all of which will even work—and gradually sculpt it into a useful device. Instead of taking the traditional top-down approach, the bottom-up approach will probably soon have to be adopted because of the sheer size of these nanocircuits. Not everything in the circuit will probably work because at the nano level, nanocircuits will be more defective and faulty because of their compactness.

Scientists and engineers have created all of the essential components of nanocircuits such as transistors, logic gates and diodes. They have all been constructed from organic molecules, carbon nanotubes and nanowire semiconductors. The only thing left to do is find a way to eliminate the errors that come with such a small device and nanocircuits will become a way of all electronics. However, eventually there will be a limit as to how small nanocircuits can become and computers and electronics will reach their equilibrium speeds.

Crossbar Latch

The cross-bar latch is a technology invented by Hewlett-Packard in October 2001, that potentially could replace transistors in some applications. Transistors are essential components of logic gates and memory cells in digital electronic systems.

The historical trend for computer systems to become smaller and more powerful over the years is mostly due to techniques for producing ever smaller transistors. The memristor devices that make up a cross bar latch potentially can be made smaller than any transistor.

Details

The crossbar was introduced by Hewlett-Packard scientists in the Journal of Applied Physics, and allows much the same functionality of transistors, except on a molecular scale. It has been seen by many as an advancement as far above the transistor as the latter was to the vacuum tube. The crossbar latch consists of a signal line crossed by two control lines. Depending on the voltages sent down the various lines, it can simulate the action of the three major logic gates: AND, OR and NOT.

The abstract of the patent is as follows:

> *"A molecular crossbar latch is provided, comprising two control wires and a signal wire that crosses the two control wires at a non-zero angle to thereby form a junction with each control wire. Each junction forms a switch and the junction has a functional dimension in nanometres. The signal wire selectively has at least two different voltage states, ranging from a 0 state to a 1 state, wherein there is an asymmetry with respect to the direction of current flow from the signal wire through one junction compared to another junction such that current flowing through one junction into (out of) the signal wire can open (close) while current flowing through the other junction out of (into) the signal wire can close (open) the switch, and wherein there is a voltage threshold for switching between an open switch and a closed switch. Further, methods are provided for latching logic values onto nanowires in a logic array, for inverting a logic value, and for restoring a voltage value of a signal in a nano-scale wire. Invented by Phillip J Kuekes."* — *U.S. patent* ***6,586,965***

Applications in Arithmetic Processing

This application was developed by Greg Snider of Hewlett-Packard and employs crossbar latches to replicate the functions of a half adder, which form the basis of modern computational architectures.

In this application a crossbar tile is formed of a layer of horizontal row wires and a layer of vertical column wires in which memristor or similar materials are sandwiched between the horizontal and vertical wire layers. Each intersection or junction of the crossbar tile is configurable to be at a high-resistance state in which there is little or no current between the horizontal and vertical wires or at a low-resistance state in which current is permitted. By setting different logic inputs A, NOT A, B, and NOT B to different row wires this configuration produces the sum and carry outputs typical for a half-adder. Connections between multiple half-adders may then be used to form full adders in accordance with conventional arithmetic architectures.

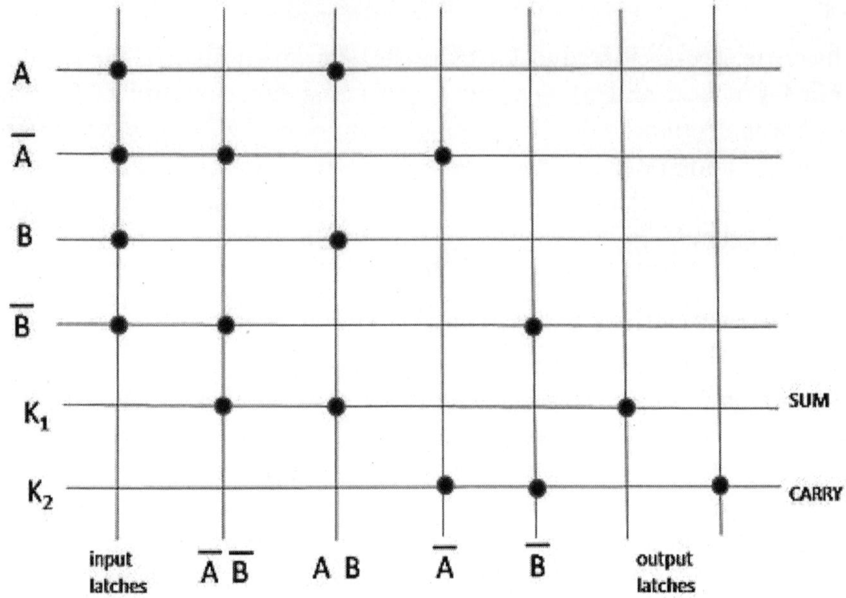

Figure: *Crossbar latches configured as a half-adder*

Applications of Crossbar Latch in Neuromorphics

Crossbar latches have been suggested as components of neuromorphic computing systems. One implementation of this is in the form of a neural network formed from nanowires as discussed in a patent by Greg Snider of Hewlett-Packard.

Memristor

The memristor (/Èm[mrhstYr/; a portmanteau of "memory resistor") was originally envisioned in 1971 by circuit theorist Leon Chua as a missing non-linear passive two-terminal electrical component relating electric charge and magnetic flux linkage. According to the governing mathematical relations, the memristor's electrical resistance is not constant but depends on the history of current that had previously flowed through the device, *i.e.*, its current resistance depends on how much electric charge has flowed in what direction through it in the past. The device remembers its history, that is, when the electric power supply is turned off, the memristor remembers its most recent resistance until it is turned on again. Thermodynamic considerations, however, show that such a memristor component cannot exist as a solid state device in physical reality because its behaviour would be inconsistent with fundamental laws of non-equilibrium thermodynamics.

Leon Chua has more recently argued that the definition could be generalized to cover all forms of 2-terminal non-volatile memory devices based on resistance switching effects although some experimental evidence contradicts this claim, since a non-passive nanobattery effect is observable in resistance switching memory. Chua also argued that the memristor is the oldest known circuit element with its effects predating the resistor, capacitor and inductor. The memristor is currently under development by various teams including Hewlett-Packard, SK Hynix and HRL Laboratories.

In 2008, a team at HP Labs claimed to have found Chua's missing memristor based on an analysis of a thin film of titanium dioxide. However, some skepticism has been expressed regarding this analysis. These devices are intended for applications in nanoelectronic memories, computer logic and neuromorphic computer architectures. In October 2011, the team announced the commercial availability of memristor technology within 18 months, as a replacement for Flash, SSD, DRAM and SRAM. Commercial availability was more recently estimated as 2018. In March 2012, a team of researchers from HRL Laboratories and the University of Michigan announced the first functioning memristor array built on a CMOS chip.

Background

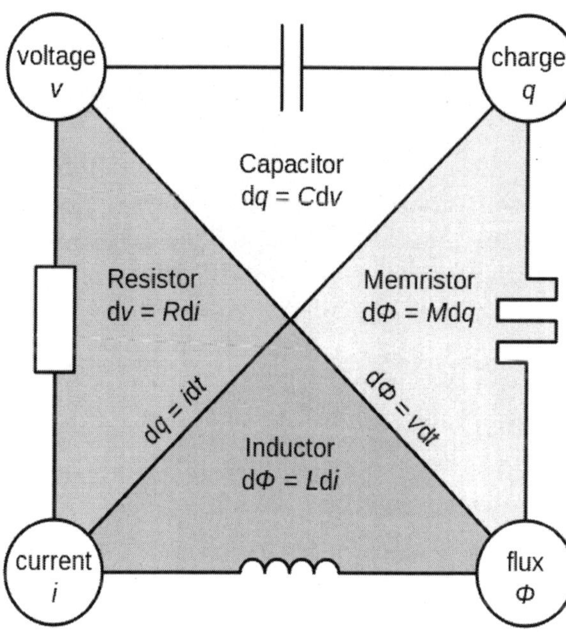

Figure: *Conceptual symmetry between the resistor, capacitor, inductor, and the memristor.*

In his 1971 paper, Chua extrapolated a conceptual symmetry between the nonlinear resistor (voltage vs. current), nonlinear capacitor (voltage vs. charge) and nonlinear inductor (magnetic flux linkage vs. current). He then inferred the possibility of a memristor as another fundamental nonlinear circuit element linking magnetic flux linkage and charge.

In contrast to a linear (or nonlinear) resistor the memristor has a dynamic relationship between current and voltage including a memory of past voltages or currents. Other scientists had proposed dynamic memory resistors such as the memistor of Bernard Widrow, but Chua attempted to introduce mathematical generality.

Memristor resistance depends on the integral of the input applied to the terminals (rather than on the instantaneous value of the input as in a varistor). Since the element "remembers" the amount of current that last passed through, it was tagged by Chua with the name "memristor". Another way of describing a memristor is as any passive two-terminal circuit element that maintains a functional relationship between the time integral of current (called charge) and the time integral of voltage (often called flux, as it is related to magnetic flux). The slope of this function is called the memristance M and is similar to variable resistance. However, purely current- or voltage-controlled memristors cannot exist in physical reality because they would then operate in conflict with fundamental laws of non-equilibrium thermodynamics.

The memristor definition is based solely on the fundamental circuit variables of current and voltage and their time-integrals, just like the resistor, capacitor and inductor. Unlike those three elements however, which are allowed in linear time-invariant or LTI system theory, memristors of interest have a dynamic function with memory and may be described as some function of net charge. There is no such thing as a standard memristor. Instead, each device implements a particular function, wherein the integral of voltage determines the integral of current, and vice versa. A linear time-invariant memristor, with a constant value for M, is simply a conventional resistor. Like other two-terminal components, real-world devices are never purely memristors ("ideal memristor"), but also exhibit some amount of capacitance, resistance and inductance.

Memristor Definition and Criticism

According to the original 1971 definition, the memristor was the fourth fundamental circuit element, forming a non-linear relationship between electric charge and magnetic flux linkage. In 2011 Chua argued for a broader definition that included all 2-terminal non-volatile memory devices based on resistance switching. Williams argued that MRAM, phase change memory and RRAM were memristor technologies. Some researchers argued that biological structures such as blood and skin fit the definition. Others argued that the memory device under development by HP Labs and other forms of RRAM were memristive systems but part of a broader class of variable resistance systems and that a broader definition of memristor is a scientifically unjustifiable land grab that favoured HP's memristor patents.

Meuffels and Schroeder noted that one of the early memristor papers included a mistaken assumption regarding ionic conduction. Meuffels and Soni discussed issues and problems in the realisation of memristors. They claimed that the physics behind the HP memristor model conflicts with fundamentals of solid state electrochemistry as the coupling of electronic/ionic diffusion currents was not considered. Additionally, they pointed to issues concerning fundamentals of non-equilibrium thermodynamics: the dynamic state equations formulated for memristors like the HP memristor imply the possibility of violating Landauer's principle of the minimum amount of energy required to change "information" states in a system. This critique was endorsed by Di Ventra and Pershin.

Nonvolatile information storage requires the existence of energy barriers that separate distinct memory states from each other. Memristors whose resistance (memory) states depend only on the current (like the HP memristor) or voltage history would be unable to protect their memory states against unavoidable fluctuations and thus permanently suffer information loss: the proposed hypothetical concept provides no physical mechanism enabling such systems to retain memory states after the applied current or voltage stress is removed. Such elements can therefore not exist, as they would always be susceptible to a so-called "stochastic catastrophe".

Other researchers noted that memristor models based on the assumption of linear ionic drift do not account for asymmetry between set time (high-to-low resistance switching) and reset time (low-to-high resistance switching) and do not provide ionic mobility values consistent with experimental data. Non-linear ionic drift models have been proposed to compensate for this deficiency.

Martin Reynolds, an electrical engineering analyst with research outfit Gartner, commented that while HP was being sloppy in calling their device a memristor, critics were being pedantic in saying it was not a memristor.

Experimental Tests for Memristors

Chua suggested experimental tests to determine if a device may properly be categorized as a memristor:

- The Lissajous curve in the voltage-current plane is a pinched hysteresis loop when driven by any bipolar periodic voltage or current without respect to initial conditions.
- The area of each lobe of the pinched hysteresis loop shrinks as the frequency of the forcing signal increases.

- As the frequency tends to infinity, the hysteresis loop degenerates to a straight line through the origin, whose slope depends on the amplitude and shape of the forcing signal.

According to Chua all resistive switching memories including ReRAM, MRAM and phase change memory meet these criteria and are memristors. However, the lack of data for the Lissajous curves over a range of initial conditions or over a range of frequencies, complicates assessments of this claim.

Experimental evidence shows that redox-based resistance memory (ReRAM) includes a nanobattery effect that is contrary to Chua's memristor model. This indicates that the memristor theory needs to be extended or corrected to enable accurate ReRAM modelling.

Theory

The memristor was originally defined in terms of a non-linear functional relationship between magnetic flux linkage $\Phi_m(t)$ and the amount of electric charge that has flowed, q(t):

$$f(\Phi_m(t), q(t)) = 0$$

The variable Φ_m ("magnetic flux linkage") is generalized from the circuit characteristic of an inductor. It *does not* represent a magnetic field here. Its physical meaning is discussed below. The symbol Φ_m may be regarded as the integral of voltage over time.

In the relationship between Φ_m and q, the derivative of one with respect to the other depends on the value of one or the other, and so each memristor is characterized by its memristance function describing the charge-dependent rate of change of flux with charge.

$$M(q) = \frac{d\Phi_m}{dq}$$

Substituting the flux as the time integral of the voltage, and charge as the time integral of current, the more convenient form is

$$M(q(t)) = \frac{d\Phi_m / dt}{dq / dt} = \frac{V(t)}{I(t)}$$

To relate the memristor to the resistor, capacitor, and inductor, it is helpful to isolate the term *M*(*q*), which characterizes the device, and write it as a differential equation.

Device	*Characteristic property (units)*	*Differential equation*
Resistor	Resistance (V per A, or Ohm, Ω)	$R = dV / dI$
Capacitor	Capacitance (C per V, or Farads)	$C = dq / dV$
Inductor	Inductance (Wb per A, or Henrys)	$L = d\Phi_m / dI$
Memristor	Memristance (Wb per C, or Ohm)	$M = d\Phi_m / dq$

The above table covers all meaningful ratios of differentials of I, Q, Φ_m, and V. No device can relate dI to dq, or $d\Phi_m$ to dV, because I is the derivative of Q and Φ_m is the integral of V.

It can be inferred from this that memristance is charge-dependent resistance. If $M(q(t))$ is a constant, then we obtain Ohm's Law $R(t) = V(t)/ I(t)$. If $M(q(t))$ is nontrivial, however, the equation is not equivalent because $q(t)$ and $M(q(t))$ can vary with time. Solving for voltage as a function of time produces

$$V(t) = M(q(t))I(t)$$

This equation reveals that memristance defines a linear relationship between current and voltage, as long as M does not vary with charge. Nonzero current implies time varying charge. Alternating current, however, may reveal the linear dependence in circuit operation by inducing a measurable voltage without net charge movement—as long as the maximum change in q does not cause much change in M.

Furthermore, the memristor is static if no current is applied. If $I(t) = 0$, we find $V(t) = 0$ and $M(t)$ is constant. This is the essence of the memory effect.

The power consumption characteristic recalls that of a resistor, I^2R.

$$P(t) = I(t)V(t) = I^2(t)M(q(t))$$

As long as $M(q(t))$ varies little, such as under alternating current, the memristor will appear as a constant resistor. If $M(q(t))$ increases rapidly, however, current and power consumption will quickly stop.

$M(q)$ is physically restricted to be positive for all values of q (assuming the device is passive and does not become superconductive at some q). A negative value would mean that it would perpetually supply energy when operated with alternating current.

In 2008 researchers from HP Labs introduced a model for a memristance function based on thin films of titanium dioxide. For $R_{ON} << R_{OFF}$ the memristance function was determined to be

$$M(q(t)) = R_{\mathrm{OFF}} \cdot \left(1 - \frac{\mu_v R_{\mathrm{ON}}}{D^2} q(t)\right)$$

where R_{OFF} represents the high resistance state, R_{ON} represents the low resistance state, μ_v represents the mobility of dopants in the thin film, and D represents the film thickness. The HP Labs group noted that "window functions" were necessary to compensate for differences between experimental measurements and their memristor model due to nonlinear ionic drift and boundary effects.

Operation as a Switch

For some memristors, applied current or voltage causes substantial change in resistance. Such devices may be characterized as switches by investigating the time and energy that must be spent to achieve a desired change in resistance. This assumes that the applied voltage remains constant. Solving for energy dissipation during a single switching event reveals that for a memristor to switch from R_{on} to R_{off} in time T_{on} to T_{off}, the charge must change by $\Delta Q = Q_{on} - Q_{off}$.

$$E_{switch} = V^2 \int_{T_{off}}^{T_{on}} \frac{dt}{M(q(t))} = V^2 \int_{Q_{off}}^{Q_{on}} \frac{dq}{I(q)M(q)} = V^2 \int_{Q_{off}}^{Q_{on}} \frac{dq}{V(q)} = V\Delta Q$$

Substituting $V = I(q)M(q)$, and then +"dq/V= "Q/V for constant V To produces the final expression. This power characteristic differs fundamentally from that of a metal oxide semiconductor transistor, which is capacitor-based. Unlike the transistor, the final state of the memristor in terms of charge does not depend on bias voltage.

The type of memristor described by Williams ceases to be ideal after switching over its entire resistance range, creating hysteresis, also called the "hard-switching regime". Another kind of switch would have a cyclic $M(q)$ so that each *off-on* event would be followed by an *on-off* event under constant bias. Such a device would act as a memristor under all conditions, but would be less practical.

Memristive Systems

The memristor was generalized to memristive systems in Chua's 1976 paper. Whereas a memristor has mathematically scalar state, a system has vector state. The number of state variables is independent of the number of terminals.

Chua applied this model to empirically-observed phenomena, including the Hodgkin–Huxley model of the axon and a thermistor at constant ambient temperature. He also described memristive systems in terms of energy storage and easily-observed electrical characteristics. These characteristics might match resistive random-access memory relating the theory to active areas of research.

In the more general concept of an n-th order memristive system the defining equations are

$$y(t) = g(\mathrm{x}, u, t)u(t),$$

$$\dot{\mathrm{x}} = f(\mathrm{x}, u, t)$$

where $u(t)$ is an input signal, $y(t)$ is an output signal, the vector x represents a set of n state variables describing the device, and g and f are continuous functions. For a current-controlled memristive system the signal *u(t)* represents the current signal, *i(t)* and the signal *y(t)* represents the voltage signal *v(t)*. For a voltage-controlled memristive system the signal *u(t)* represents the voltage signal *v(t)* and the signal *y(t)* represents the current signal *i(t)*.

The *pure* memristor is a particular case of these equations, namely when *x* depends only on charge ($x=q$) and since the charge is related to the current via the time derivative $dq/dt=i(t)$. Thus for *pure* memristors *f* (i.e. the rate of change of the state) must be equal or proportional to the current *i(t)* .

Pinched Hysteresis

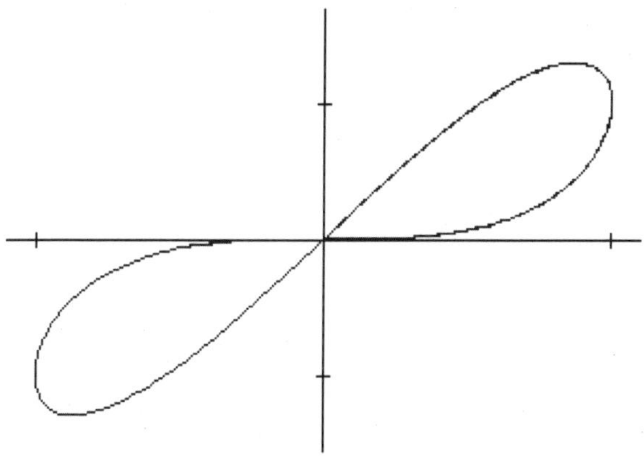

Figure: *Example of pinched hysteresis curve, V versus I*

One of the resulting properties of memristors and memristive systems is the existence of a pinched hysteresis effect. For a current-controlled memristive system, the input *u(t)* is the current *i(t)*, the output *y(t)* is the voltage *v(t)*, and the slope of the curve represents the electrical resistance. The change in slope of the pinched hysteresis curves demonstrates switching between different resistance states which is a phenomenon central to ReRAM and other forms of two-terminal resistance memory. At high frequencies, memristive theory predicts the pinched hysteresis effect will degenerate, resulting in a straight line representative of a linear resistor. It has been proven that some types of non-crossing pinched hysteresis curves (denoted Type-II) cannot be described by memristors.

Extended Memristive Systems

Some researchers have raised the question of the scientific legitimacy of HP's memristor models in explaining the behaviour of ReRAM. and have suggested extended memristive models to remedy perceived deficiencies.

One example attempts to extend the memristive systems framework by including dynamic systems incorporating higher-order derivatives of the input signal *u(t)* as a series expansion

$$y(t) = g_0(\mathrm{x},u)u(t) + g_1(\mathrm{x},u)\frac{\mathrm{d}^2 u}{\mathrm{d}t^2} + g_2(\mathrm{x},u)\frac{\mathrm{d}^4 u}{\mathrm{d}t^4} + \ldots + g_m(\mathrm{x},u)\frac{\mathrm{d}^{2m} u}{\mathrm{d}t^{2m}},$$

$$\dot{\mathrm{x}} = f(\mathrm{x},u)$$

where m is a positive integer, $u(t)$ is an input signal, $y(t)$ is an output signal, the vector x represents a set of n state variables describing the device, and the functions g and f are continuous functions. This equation produces the same zero-crossing hysteresis curves as memristive systems but with a different frequency response than that predicted by memristive systems.

Another example suggests including an offset value a to account for an observed nanobattery effect which violates the predicted zero-crossing pinched hysteresis effect.

$$y(t) = g_0(\mathrm{x},u)(u(t) - a),$$

Implementations

Titanium Dioxide Memristor: Interest in the memristor revived when an experimental solid state version was reported by R. Stanley Williams of Hewlett Packard. The article was the first to demonstrate that a solid-state device could have the characteristics of a memristor based on the behaviour of nanoscale thin films. The device neither uses magnetic flux as the theoretical memristor suggested, nor stores charge as a capacitor does, but instead achieves a resistance dependent on the history of current.

Although not cited in HP's initial reports on their TiO_2 memristor, the resistance switching characteristics of titanium dioxide were originally described in the 1960s.

The HP device is composed of a thin (50 nm) titanium dioxide film between two 5 nm thick electrodes, one titanium, the other platinum. Initially, there are two layers to the titanium dioxide film, one of which has a slight depletion of oxygen atoms. The oxygen vacancies act as charge carriers, meaning that the depleted layer has a much lower resistance than the non-depleted layer. When an electric field is applied, the oxygen vacancies drift, changing the boundary between the high-resistance and low-resistance layers. Thus the resistance of the film as a whole is dependent on how much charge has been passed through it in a particular direction, which is reversible by changing the direction of current. Since the HP device displays fast ion conduction at nanoscale, it is considered a nanoionic device.

Memristance is displayed only when both the doped layer and depleted layer contribute to resistance. When enough charge has passed through the memristor that the ions can no longer move, the device enters hysteresis. It ceases to integrate $q=+\int I dt$, but rather keeps q at an upper bound and M fixed, thus acting as a constant resistor until current is reversed.

Memory applications of thin-film oxides had been an area of active investigation for some time. IBM published an article in 2000 regarding structures similar to that described by Williams. Samsung has a U.S. patent for oxide-vacancy based switches similar to that described by Williams. Williams also has a U.S. patent application related to the memristor construction.

In April 2010, HP labs announced that they had practical memristors working at 1 ns (~1 GHz) switching times and 3 nm by 3 nm sizes, which bodes well for the future of the technology. At these densities it could easily rival the current sub-25 nm flash memory technology.

Polymeric Memristor

In 2004, Krieger and Spitzer described dynamic doping of polymer and inorganic dielectric-like materials that improved the switching characteristics and retention required to create functioning nonvolatile memory cells. They used a passive layer between electrode and active thin films, which enhanced the extraction of ions from the electrode. It is possible to use fast ion conductor as this passive layer, which allows a significant reduction of the ionic extraction field.

In July 2008, Erokhin and Fontana claimed to have developed a polymeric memristor before the more recently announced titanium dioxide memristor.

In 2012, Crupi, Pradhan and Tozer described a proof of concept design to create neural synaptic memory circuits using organic ion-based memristors. The synapse circuit demonstrated long-term potentiation for learning as well as inactivity based forgetting. Using a grid of circuits, a pattern of light was stored and later recalled. This mimics the behaviour of the V1 neurons in the primary visual cortex that act as spatiotemporal filters that process visual signals such as edges and moving lines.

Ferroelectric Memristor

The ferroelectric memristor is based on a thin ferroelectric barrier sandwiched between two metallic electrodes. Switching the polarization of the ferroelectric material by applying a positive or negative voltage across the junction can lead to a two order of magnitude resistance variation: $R_{OFF} >> R_{ON}$ (an effect called Tunnel Electro-Resistance). In general, the polarization does not switch abruptly. The reversal occurs gradually through the nucleation and growth of ferroelectric domains with opposite polarization. During this process, the resistance is neither R_{ON} or R_{OFF}, but in between. When the voltage is cycled, the ferroelectric domain configuration evolves, allowing a fine tuning of the resistance value. The ferroelectric memristor's main advantages are that ferroelectric domain dynamics can be tuned, offering a way to engineer the memristor response, and that the resistance variations are due to purely electronic phenomena, aiding device reliability, as no deep change to the material structure is involved.

Spin Memristive Systems

Spintronic Memristor

Chen and Wang, researchers at disk-drive manufacturer Seagate Technology described three examples of possible magnetic memristors. In one device resistance occurs when the spin of electrons in one section of the device points in a different direction from those in another section, creating a "domain wall", a boundary between the two sections.

Electrons flowing into the device have a certain spin, which alters the device's magnetization state. Changing the magnetization, in turn, moves the domain wall and changes the resistance. The work's significance led to an interview by IEEE Spectrum. A first experimental proof of the spintronic memristor based on domain wall motion by spin currents in a magnetic tunnel junction was given in 2011.

Spin-Transfer Torque Magnetoresistance

Spin-transfer torque MRAM is a well-known device that exhibits memristive behaviour. The resistance is dependent on the magnetic state of a magnetic tunnel junction, i.e., on the relative magnetization alignment of the two electrodes. This can be controlled by spin torque induced by current flowing through the junction. However, the length of time the current flows through the junction determines the amount of current needed, i.e., charge is the key variable.

Additionally, Krzysteczko et al., reported that MgO-based magnetic tunnel junction s show memristive behaviour based on the drift of oxygen vacancies within the insulating MgO layer (resistive switching).

Therefore, the combination of spin-transfer torque and resistive switching leads naturally to a second-order memristive system described by the state vector $x=(x_1,x_2)$, where x_1 describes the magnetic state of the electrodes and x_2 denotes the resistive state of the MgO barrier. In this case the change of x_1 is current-controlled (spin torque is due to a high current density) whereas the change of x_2 is voltage-controlled (the drift of oxygen vacancies is due to high electric fields). The presence of both effects in a memristive magnetic tunnel junction led to the idea of a nanoscopic synapse-neuron system.

Spin Memristive System

A fundamentally different mechanism for memristive behaviour has been proposed by Pershin and Di Ventra. The authors show that certain types of semiconductor spintronic structures belong to a broad class of memristive systems as defined by Chua and Kang.

The mechanism of memristive behaviour in such structures is based entirely on the electron spin degree of freedom which allows for a more convenient control than the ionic transport in nanostructures. When an external control parameter (such as voltage) is changed, the adjustment of electron spin polarization is delayed because of the diffusion and relaxation processes causing hysteresis. This result was anticipated in the study of spin extraction at semiconductor/ferromagnet interfaces, but was not described in terms of memristive behaviour. On a short time scale, these structures behave almost as an ideal memristor. This result broadens the possible range of applications of semiconductor spintronics and makes a step forward in future practical applications.

Applications

Williams' solid-state memristors can be combined into devices called crossbar latches, which could replace transistors in future computers, given their much higher circuit density.

They can potentially be fashioned into non-volatile solid-state memory, which would allow greater data density than hard drives with access times similar to DRAM, replacing both components. HP prototyped a crossbar latch memory that can fit 100 gigabits in a square centimetre, and proposed a scalable 3D design (consisting of up to 1000 layers or 1 petabit per cm^2). In May 2008 HP reported that its device reaches currently about one-tenth the speed of DRAM. The devices' resistance would be read with alternating current so that the stored value would not be affected. In May 2012 it was reported that access time had been improved to 90 nanoseconds if not faster, approximately one hundred times faster than contemporaneous flash memory, while using one percent as much energy.

Memristor patents include applications in programmable logic, signal processing, neural networks, control systems, reconfigurable computing, brain-computer interfaces and RFID. Memristive devices are potentially used for stateful logic implication, allowing a replacement for CMOS-based logic computation. Several early works in this direction are reported.

In 2009, a simple electronic circuit consisting of an LC network and a memristor was used to model experiments on adaptive behaviour of unicellular organisms. It was shown that subjected to a train of periodic pulses, the circuit learns and anticipates the next pulse similar to the behaviour of slime molds Physarum polycephalum where the viscosity of channels in the cytoplasm responds to periodic environment changes. Applications of such circuits may include, e.g., pattern recognition. The DARPA SyNAPSE project funded HP Labs, in collaboration with the Boston University Neuromorphics Lab, to develop neuromorphic architectures which may be based on memristive systems. In 2010, Versace and Chandler described the MoNETA (Modular Neural Exploring Travelling Agent) model. MoNETA is the first large-scale neural network model to implement whole-brain circuits to power a virtual and robotic agent using memristive hardware. Application of the memristor crossbar structure in the construction of an analog soft computing system was demonstrated by Merrikh-Bayat and Shouraki. In 2011 they showed how memristor crossbars can be combined with fuzzy logic to create an analog memristive neuro-fuzzy computing system with fuzzy input and output terminals. Learning is based on the creation of fuzzy relations inspired from Hebbian learning rule.

In 2013 Leon Chua published a tutorial underlining the broad span of complex phenomena and applications that memristors span and how they can be used as non-volatile analog memories and can mimic classic habituation and learning phenomena.

Memcapacitors and Meminductors

In 2009, Di Ventra, Pershin and Chua extended the notion of memristive systems to capacitive and inductive elements in the form of memcapacitors and meminductors, whose properties depend on the state and history of the system, further extended in 2013 by Di Ventra and Pershin.

Bibliography

Altmann, J. & Gubrud, M.A.: *Discovering the Nanoscale*, IOS Press, Amsterdam, 2004.

Andrew, N.: *Foundations of Nanomechanics: from Solid-state Theory to Device Applications*, Berlin, New York, 2003.

Attebery, B.: *Nanoculture: Implications of the New Technoscience*, Intellect Books, Bristol, 2004.

Bainbridge, W.S. : *Societal Implications of Nanoscience and Nanotechnology*, Dordrecht, Kluwer, 2001.

Bast, R.C. Jr.: *Cancer Medicine*, Decker, Hamilton, 2000.

Berube, D.M.: *Discovering the Nanoscale*, IOS Press, Amsterdam, 2004.

Brosnan, J.: *The Primal Scream: A History of Science Fiction Film*, Orbit Books, London, 1991.

Carroll, J.S.: *Social Science Research Methods for Assessing Societal Implications of Nanotechnology*, Dordrecht, Kluwer, 2001.

Caviedes, C.: *El Nino in History: Storming Through the Ages*, University Press of Florida, Florida, 2001.

Credi, A.: *Molecular Devices and Machines: a Journey into the Nano World*, Wiley-VCH, Weinheim, 2003.

Crow, M. & Sarewitz, D.: *Societal Implications of Nanoscience and Nanotechnology*, Dordrecht: Kluwer, 2001.

Daniel Ratner: *Nanotechnology: a Gentle Introduction to the Next Big Idea*, Upper Saddle River, Prentice Hall, 2003.

Dobzhanshy, T.: *Genetics of the Evolutionary Process*, Columbia University Press, New York, 1970.

Drexler, K. Eric: *Engines of Creation: The Coming Era of Nanotechnology*, Anchor Books, New York, 1986.

Dyson, Tim: *India's Historical Demography: Studies in Famine, Disease, and Society*, Riverdale, Westwood, MA, 1989.

Eamonn Gearon: *The Sahara: A Cultural History*, Signal Books, UK, 2011.

Frank J. Owens: *Introduction to Nanotechnology*, Hoboken, J. Wiley, 2003.

Glimell, H.: *Dynamics of the Emerging Field of Nanoscience*, Dordrecht, Kluwer, 2001.

Glimell, H.: *Societal Implications of Nanoscience and Nanotechnology*, Kluwer, Dordrecht, 2001.

Haber, L. F.: *The Chemical Industry: 1900-1930, International Growth and Technological Change*, Oxford: Claredon Press, 1971.

Hessenbruch, A.: *Nanotechnology and the Negotiation of Novelty*, Amsterdam, IOS Press, 2004.

James Lewis: *Prospects in Nanotechnology: Toward Molecular Manufacturing,* John Wiley & Sons, New York, 1995.

Karn, Barbara: *Nanotechnology and the Environment: Applications and Implications,* American Chemical Society, Washington D.C., 2004.

Keblinski, P. : *Encyclopedia of Nanoscience and Nanotechnology*, American Scientific, Los Angeles, Calif, USA, 2004.

Liu W.K., Karpov E.G., Park H.S.: *Nano Mechanics and Materials,* Wiley, Weinheim, 2006.

Mayr, E.: *Change of Genetic Environment and Evolution*, Allen and Unwin, London, 1954.

Mody, C.C.M.: *How Probe Microscopists Became Nanotechnologists, Discovering the Nanoscale*, IOS Press, Amsterdam, 2004.

Norio Taniguchi: *Nanotechnology: Integrated Processing Systems for Ultra-precision and Ultra-fine Products,* Oxford University Press, New York, 1996.

Orth, R. N.: *Nanometer Scale Antibody Patterning for Directed cell Immobilization and Stimulation,* Cornell University, Ithaca, 2002.

Pressman, J.: *Nano Narrative: A Parable from Electronic Literature*, Intellect Books, Bristol, 2004.

Pressman, J.: *Nanoculture: Implications of the New Technoscience*, Intellect Books, Bristol, UK, 2004.

Ratner, Daniel: *Nanotechnology: a Gentle Introduction to the Next Big Idea*, Prentice Hall, Upper Saddle River, 2003.

Robison, W.L.: *Discovering the Nanoscale*, IOS Press, Amsterdam, 2004.

Sarewitz, D.: *Nanotechnology and Societal Transformation*, Dordrecht, Kluwer, 2001.

Stephen M. Goodnick: *Transport in Nanostructures,* Cambridge University Press, Cambridge, 1997.

Tenner, E.: *Nanotechnology and Unintended Consequences*, , Dordrecht, Kluwer, 2001.

Vasudev Bhatnagar: *Cell Science and Technology*, Campus Books, Delhi, 2009.

Vermaas, P.E.: *Discovering the Nanoscale*, IOS Press, Amsterdam, 2004.

Weil, V.: *Ethical Issues in Nanotechnology*, Dordrecht, Kluwer, 2001.

Index

L

M

N

O

P

Q

R

S

T

U

V

X

❑❑❑